Large Deviations Applied to Classical and Quantum Field Theory

Large Deviations Applied to Classical and Quantum Field Theory

Harish Parthasarathy
Professor
Electronics & Communication Engineering
Netaji Subhas Institute of Technology (NSIT)
New Delhi, Delhi-110078

CRC Press
Taylor & Francis Group
Boca Raton London New York

CRC Press is an imprint of the
Taylor & Francis Group, an informa business

Manakin
PRESS

First published 2023
by CRC Press
4 Park Square, Milton Park, Abingdon, Oxon, OX14 4RN

and by CRC Press
6000 Broken Sound Parkway NW, Suite 300, Boca Raton, FL 33487-2742

CRC Press is an imprint of Informa UK Limited

British Library Cataloguing-in-Publication Data
A catalogue record for this book is available from the British Library

ISBN: 9781032425474 (hbk)
ISBN: 9781032425498 (pbk)
ISBN: 9781003363248 (ebk)

DOI: 10.4324/9781003363248

Typeset in Arial, Minion Pro, Times New Roman, Rupee, Wingdings, Calibri, Symbol
by Manakin Press, Delhi

Manakin
PRESS

Preface

This book deals with a variety of problems in physics and engineering where the Large deviation principle of probability finds application. Large deviations is a branch of probability theory dealing with approximate computation of the probabilities of rare events. Specifically we have a sequence of process valued random variables which converges to zero or to a constant as the parameter that indexes the process converges to zero or to some other constant. Physically this means that the parameter defines the degree of noise in the system that generates the process valued r.v. and in the limit when this parameter converges to zero, the noise amplitude also converges to zero. In such a situation, the probability distribution of the process becomes degenerate in the limit as for example happens in the weak law of large numbers. We are then interested in the rate at which this probability distribution converges to the degenerate distribution. Such a calculation then gives us the asymptotic probability of the rate event that the process valued r.v. will assume values in a set that does not contain the degenerate limit. The expression for this asymptotic probability of deviation is determined by its rate function and is usually a much simpler expression than that for the exact probability of deviation. Therefore we can use this rate function to control the system parameters so that the deviation probability is minimized. This is one application of the LDP to control theory. Other applications deal with electromagnetism, quantum mechanics, general relativity, mechanics, cosmology, quantum field theory and quantum stochastic processes and string theory which is the modern theory of quantum gravity. This book analyzes a variety of such problems where the LDP can be applied. For example in quantum mechanics and quantum field theory, we have an electron bound to its nucleus on which a random incident electromagnetic radiation field is incident. When the random noise component in the field is small then we can compute the effect of this component on the transition probability of the electron between two of its stationary states and control the parameters in the non-random field component so that this change in the transition probability is a minimum. Such a computation would enable us to design a monochromatic laser that is nearly insensitive to noisy perturbations. In general relativity, we have for example the problem of controlling the background metric so that the influence of random electromagnetic sources on the nature of gravitational waves propagating in the background is minimized. This would enable us to design more and more accurate tests for Einstein's general theory of relativity. In quantum field theory, we have for example the problem of controlling the electron and photon propagators by applying external classical fields and currents to a system so that the effect of random noise in the environment on the propagator deviations is minimized. This effect is measured by the probability of deviation of the propagator function from the desired one in the absence of the environmental noise. This would enable us to design more robust accelerators for studying the S-matrix involving scattering, absorption and emission of elementary particles. This book talks about several such examples. It contains application s of the LDP to pattern recognition problems like for example analysis of the performance of the EM algorithm for optimal parameter estimation in the presence of weak noise, analysis and control of non-Abelian gauge fields in the presence of noise, quantum gravity wherein we are concerned with perturbation to the quadratic component of the Einstein-Hilbert Hamiltonian caused by higher order nonlinear terms in the position fields and their effect on the Gibbs statistics and consequently quantum probabilities of events computed using the quantum Gibbs state. The reader will also find in this book application s of LDP to quantum filtering theory as developed by Belavkin based on the celebrated Hudson-Parthasarathy quantum stochastic calculus. The idea here is that the estimate of the system density operator based on non-demolition measurements follows an Abelian and hence classical stochastic dynamics and if the Lindblad noise parameters in the HP equation are small, then we can in principle compute the probability distribution of the filtered/estimated state process using its rate function in the limit of zero Lindblad noise parameters. Applications to string theory involve computing the change in the action functional of a point field caused by its string theoretic evaluation followed by quantum averaging with respect to the quantum fluctuating part of the string in a coherent state. This idea also gives us a method to obtain string theoretic corrections to point field theories. The book will be of use to graduate students in engineering and mathematical physics as well as researchers in these fields.

Author

Brief Contents

Detailed Contents

Chapter 1

LDP Problems in Quantum Field Theory

1.1 Large deviations for Supergravity fields

Consider the supergravity Lagrangian

$$L = e.R + C_1\bar{\chi}_\mu\Gamma^{\mu\nu\rho}D_\nu\chi_\rho$$

where $e = \sqrt{-g}$, $R = R^{mn}_{\mu\nu}e^\mu_m e^\nu_n$,

$$R^{mn}_{\mu\nu} = \omega^{mn}_{\nu,\mu} - \omega^{mn}_{\mu,\nu} - [\omega_\mu, \omega_\nu]^{mn}$$

is the spinor representation of the gravitational curvature tensor, ie,

$$D_\mu = \partial_\mu + \omega^{mn}_\mu\Gamma_{mn}$$

and

$$R_{\mu\nu} = [D_\mu, D_\nu] = R^{mn}_{\mu\nu}\Gamma_{mn},$$

$$\Gamma_m e^m_\mu = \Gamma_\mu, \Gamma^m e^\mu_m = \Gamma^\mu$$

The formula for the spinor connection ω^{mn}_μ of the gravitational field is obtained from the variational equation

$$\delta_\omega \int L d^D x = 0$$

To derive this, we first observe that

$$\delta_\omega R = -\delta\omega^{mn}_\nu(e^\mu_m e^\nu_n)_{,\mu}$$

$$+\delta\omega^{mn}_\mu(e^\mu_m e^\nu_n)_{,\nu} + ([\delta\omega_\nu, \omega_\mu]^{mn}e^\mu_m e^\nu_n + [\omega_\nu, \delta\omega_\mu]^{mn})e^\mu_m e^\nu_n$$

The coefficient of $\delta\omega^{mn}_\mu$ is extracted as

$$-(e^\nu_m e^\mu_n)_{,\nu} + (e^\mu_m e^\nu_n)_{,\nu}$$

+

In the absence of the gravitino, we see that ω^{μ}_{mn} satisfies

$$0 = D_\mu e^m_\nu = e^m_{\nu,\mu} - \Gamma^\rho_{\mu\nu} e^m_\rho + \omega^{mn}_\mu e^n_\nu = 0$$

This equation is a linear algebraic equation for ω and is immediately solved. In the presence of the gravitino field, the solution for ω has an additional bilinear term in the gravitino field. Note that the supergravity action written above is invariant under the local supersymmetry transformations

$$\delta\chi_\mu = D_\mu \epsilon, \delta e^n_\mu = \bar{\epsilon}\Gamma^n \chi_\mu$$

where $\epsilon(x)$ is an infinitesimal Majorana Fermionic field.

The large deviation problem: Write down the supersymmetric field equations for the graviton tetrad e^m_μ and the gravitino χ_μ. Then, break this supersymmetry infintesimally by a small random perturbation Lagrangian in the tetrad and the gravitino field and for the resulting solution, calculate the Large deviation rate function using which determine approximately the probability that the tetrad and gravitino fields will deviate from the original nonrandom supersymmetric solution by an amount greater than a given threshold over a given region of space-time.

1.2 Rate function of string propagator

Consider the Bosonic vertex operator in string theory

$$V(k, z) =: exp(ik.X(z)) :$$

where : denotes normal ordering. Here,

$$X^\mu(z) = -i \sum_{n \neq 0} \alpha^\mu(n) z^n / n - ip^\mu log(z) + x^\mu$$

Note that we can regard $z = exp(i(\tau - \sigma))$ and then

$$X^\mu(z) = x^\mu + p^\mu(\tau - \sigma) + \sum_{n \neq 0} \alpha^\mu(n) exp(in(\tau - \sigma))/n$$

which is the general solution to the string differential equation

$$(\partial_\tau^2 - \partial_\sigma^2) X^\mu = 0$$

subject to the condition that X^μ is periodic in σ with a period of 2π. Then normal ordering gives

$$V(k, z) = exp(ik.x + k.p.log(z)).exp(-i \sum_{n<0} k.\alpha(n) z^n / n).exp(-i \sum_{n>0} k.\alpha(n) z^n / n)$$

where

$$k.\alpha(n) = k_\mu \alpha^\mu(n)$$

Recall the Bosonic string commutation relations

$$[\alpha^\mu(n), \alpha^\nu(n)] = n\eta^{\mu\nu}\delta[n+m]$$

Also note that

$$[x^\mu, p^\nu] = i\eta^{\mu\nu}$$

and x^μ, p^μ commute with all the $\alpha^\mu(n)'s$. Using the fact that if A, B are operators in a Hilbert space such that $C = [A, B]$ commutes with both A and B, then

$$exp(A+B) = exp(C/2).exp(A).exp(B)$$

evaluate the vertex operator commutation relations

$$[V(k_1, z_1), V(k_2, z_2)]$$

and hence evaluate

$$-\int V(k_1, z_1), V(k_2, z_2)]dz_1 dz_2/z_1 z_2 = -[\int V(k_1, z)dz/z, \int V(k_2, z)dz/z]$$

$$= [\int_0^{2\pi} V(k_1, exp(i\tau))d\tau, \int_0^{2\pi} V(k_2, exp(i\tau))d\tau]$$

where the contour for the z_1, z_2 integrals is taken to be the unit circle.

Evaluate then the string scattering amplitude with M interactions:

$$A_M(1, 2, ..., N|f, i) =< f|\Delta.V(1)\Delta.V(2)..V(M)\Delta|i >$$

where $|i>$ is the initial state, $|f>$ is the final state, $V(m) = V(k_m, z_m)$,

$$\Delta = (L_0 - 1)^{-1} = \int_0^1 z^{L_0 - 2} dz$$

is the string propagator where L_0 is the string Hamiltonian defined by

$$L_0 = p^2 + \sum_{n \geq 1} \alpha(-n).\alpha(n)$$

with

$$a.b = \eta_{\mu\nu} a^\mu b^\nu$$

Note that a large class of string observables can expressed in terms of the Vertex operators as

$$Y(\tau) = \int V(k, z) f(k) d^D k d\sigma, z = exp(i(\tau - \sigma))$$

where $f(k, z)$ is a complex valued function of (k, z). Thus, using the above string amplitude, we can evaluate things like

$$< f|U(\infty, \tau_N)Y(\tau_N)U(\tau_N, \tau_{N-1})Y(\tau_{N-1})...Y(\tau_1)U(\tau_1, -\infty)|i > --(1)$$

where

$$U(t, \tau) = exp(-i(t-\tau)L_0) = w^{L_0}, w = exp(-i(t-\tau))$$

is the free string evolution operator from time τ upto time t. (1) describes the amplitude of scattering of the string from an initial state $|i>$ at time $-\infty$ to a final state $|f>$ at time $+\infty$, when the string successively interacts with an interaction Hamiltonian $Y(t)$ at times $\tau_1, ..., \tau_N$. Note that the Heisenberg operator evolution dynamics states that

$$exp(i\tau L_0)V(k, 1).exp(-i\tau L_0) = V(k, exp(i\tau))$$

since the string Heisenberg dynamics can be expressed as

$$exp(i\tau L_0).X^\mu(1)exp(-i\tau L_0) = X^\mu(z), z = exp(i\tau)$$

Note that

$$\Delta.V(k, z).\Delta = \int_0^1 w^{L_0-2}.V(k, z).\int_0^1 w^{L_0-2}dw$$

$$= \int_0^1\int_0^1 w_1^{L_0-2}V(k, z)w_2^{L_0-2}dw_1dw_2$$

and

$$w_1^{L_0-2}V(k, z).w_2^{L_0-2} = w_1^{L_0-2}V(k, z)w_1^{-(L_0-2)}.(w_2/w_1)^{L_0-2}$$

$$= V(k, w_1z).(w_2/w_1)^{L_0-2}$$

and likewise,

$$w_1^{L_0-2}V(k, z).w_2^{L_0-2} = (w_1/w_2)^{L_0-2}w_2^{-(L_0-2)}V(k, z).w_2^{L_0-2}$$

$$= (w_1/w_2)^{L_0-2}V(k, z/w_2)$$

Now suppose that the string interacts with an external gauge field $H_{\mu\nu\rho}$ derived from a antisymmetric gauge potential $B_{\mu\nu}$, ie

$$H = dB, B = B_{\mu\nu}dX^\mu\wedge dX^\nu, H = B_{\mu\nu,\sigma}dX^\sigma\wedge dX^\mu\wedge dX^\nu = H_{\sigma\mu\nu}dX^\sigma\wedge dX^\mu\wedge d$$

or equivalently,

$$H_{\mu\nu\sigma} = B_{\mu\nu,\sigma} + B_{\nu\sigma,\mu} + B_{\sigma\mu\nu}$$

Then what corrections does the string propagator acquire due to perturbations by this gauge field ? For that, we must first write down the field equations for the string in the presence of this external gauge fields and then using these field equations, derive a differential equation for the propagator and solve it approximately under the weak gauge field assumption. The field equations have the form

$$\Box X^{\mu} = K.\epsilon^{\alpha\beta}(B^{\mu}_{\nu}(X)X^{\nu}_{,\alpha})_{,\beta}$$

where

$$\Box = \partial^{\alpha}\partial_{\alpha}$$

The LDP problem: If the external gauge field is a weak Gaussian field, then what is the rate function for the string propagator? Using this rate function, can we compute the approximate probability that the scattering amplitude for a given process will fall in a given range ? If we propose to design a quantum gate based on such a scattering process, then we can use this computation to predict the approximate probability that in the presence of such a small random gauge field, the designed gate will differ from the gate designed in the absence of the gauge field by an amount greater than a given threshold. In other words, we are regarding the gauge field as some sort of noise and we are trying to assess the performance of our designed gate in the presence of such noise.

1.3 Large deviations for p-form fields

If $C(x)$ is an r-form on an N-dimensional differentiable manifold $\mathcal{M}$, then we can write

$$C(x) = C_{i_1 \ldots i_r}(x) dx^{i_1} \wedge \ldots \wedge dx^{i_r}$$

and then we can formulate an action

$$S[C] = \int_M < dC, dC > -K \int < C, C >= \int_M (dC \wedge *dC - KC \wedge *C)$$

where if A, B are two p forms,

$$< A, B >= \epsilon(i_1, ..., i_N) A_{i_1 \ldots i_p}(x)(*B)_{i_{p+1},\ldots,i_N}(x)$$

where ϵ is the totally antisymmetric tensor in N indices and $*B$ is the antisymmetric tensor with $N - p$ indices defined by

$$(*B)_{i_{p+1}\ldots i_N} = B_{i_1 \ldots i_p}(x)$$

where $(i_{p+1}, ..., i_N)$ is the complement of the set $(i_1, ..., i_p)$ in $(1, 2, ..., N)$. Note that

$$\int < A, B > d^N x = \int_M A \wedge *B$$

The field equations for C can be derived from the variational principle

$$\delta S[C] = 0$$

Note that if A is a p-form and B a $p + 1$ form, then

$$\int < dA, B >= \int < A, *d * B >$$

which is equivalent to saying that the adjoint of d is $*d*$. We write this relation as

$$d^* = *d*$$

For example, for the Maxwell tensor $F = F_{\mu\nu} dx^\mu \wedge dx^\nu$, we have

$$(*F)^{\mu\nu} = \epsilon^{\mu\nu\rho\sigma} F_{\rho\sigma}$$

and hence

$$\int F \wedge *F = \int F_{\mu\nu} F^{\mu\nu} d^4x$$

gives the action for the electromagnetic field. The Maxwell equations in the absence of charges and currents are $d * F = 0$ or equivalently $*d * F = 0$. To check this, we observe that

$$*F = \epsilon(\mu\nu\rho\sigma) F^{\rho\sigma} dx^\mu \wedge dx^\nu$$

and hence,

$$d * F = \epsilon(\mu\nu\rho\sigma) F^{\rho\sigma}_{,\alpha} dx^\alpha \wedge dx^\mu \wedge dx^\nu$$

and the vanishing of this is equivalent to the vanishing of

$$\epsilon(\mu\nu\rho\sigma)\epsilon(\beta\alpha\mu\nu) F^{\rho\sigma}_{,\alpha}$$

which is the same as $F^{\beta\alpha}_{,\alpha}$.

The Large deviation problem: Suppose we add to the above action a small term coming from random forcing so that the action integral becomes

$$S[C] + \delta S[C] = \int < dC, dC > -K \int < C, C > +\delta. \int < f, C >$$

$$= \int dC \wedge *dC - K \int C \wedge *C + \delta \int f \wedge *C$$

where C is a p-form and f is a random noisy p-form. Then, if f is a zero mean Gaussian random field, the problem is to write down the field equations as

$$d^* dC + KC + \delta f = 0$$

and then calculate the rate function for C in terms of the correlation function of f.

Remark: In the case of the electromagnetic field, the above equation with $C = A$ so that $F = dA$ becomes

$$\Box A^\mu + KA^\mu + \delta.f^\mu = 0$$

and this equation actually represents a photon field in which the photon has a mass determined by K subject to a random current field f^μ. Note that such situation occurs in the electroweak theory where A^μ actually represents a Lie algebra field, namely the potential of the gauge bosons and the mass terms arise because of the interaction of this gauge field with the scalar Higgs field followed by symmetry breaking in which the Higgs field falls into the ground state. The above equation can formally be solved using N-dimensional Fourier integrals as

$$A^{\mu}(x) = \delta. \int D(x - x'|K) f^{\mu}(x') d^N x'$$

$$D(x|K) = -\int exp(ik.x) d^N k/(K - k^2)$$

Problem: Show that

$$\int < A, dB >= \int < d^* A, B >, d^* = *d*$$

where A is a p-form and B is a $p-1$ form. Do this problem using integration by parts.

1.4 The dynamics of the electro-weak theory

$$L(\psi, A, \phi) = \bar{\psi}(\gamma^{\mu}(i\partial_{\mu} + eA_{\mu}))\psi$$
$$+|(i\partial_{\mu} + eA_{\mu})\phi|^2 + Re(g^T(\psi \otimes \psi^* \otimes \phi))$$

where g is a coupling constant vector, ϕ is the scalar Higgs doublet field ψ is the electron-Lepton field and A is the gauge boson-photon field. Symmetry breaking of ϕ when it falls into the ground state gives a mass term $A_{\mu}A^{\mu}$ to the gauge boson field, via the second term in this Lagrangian, symmetry breaking of ϕ also gives a mass term $\psi \times \psi^*$ to the electron via the last term. Now we analyze the large deviation problem originating from this. Suppose that we add to this Lagrangian, a random external classical Hadronic current J^{μ} that interacts with the gauge boson field A_{μ} via the interaction Lagrangian $J^{\mu u}A^u_{\mu}$. Suppose that we also add to this Lagrangian an interaction term $g_1 A^c_{\mu}\bar{\psi}\gamma^{\mu}\psi$ between the electron-Lepton current an an external classical electromagnetic field A^c_{μ}. Suppose that finally, we also add to this Lagrangian an interaction term $A^c_{\mu} Im(\phi^* \partial_{\mu}\phi)$ between the Higgs scalar current and the classical external electromagnetic field. Then, we can ask that in the limit as the coupling constants in these interaction terms become very small, what is the LDP rate function for the quantum average of an observable built out of the wave field operators ψ, A_{μ}, ϕ ?

1.5 Filtering in Fermionic noise

$$dJ = (-1)^{\Lambda} dA, dJ^* = (-1)^{\Lambda} dA^*$$

$$dJdJ^* = dt, dJdA^* = (-1)^{\Lambda} dt, dAdJ^* = (-1)^{\Lambda} dt,$$

$$A(s)dJ(t) = -dJ(t)A(s), s \leq t$$

HP qsde

$$dU = (-(iH+P)dt + L_1 dA + L_2 dA^* + L_3 dJ + L_4 dJ^*)U$$

$$dU^*U + U^*dU + dU^*dU = d(U^*U) = 0$$

gives

$$0 = U^*((iH-P)dt + L_1^* dA^* + L_2^* dA + L_3^* dJ^*$$

$$+L_4^* dJ - (iH+P)dt + L_1 dA + L_2 dA^* + L_3 dJ + L_4 dJ^*)U$$

$$+U^*(L_1^d A^* + L_2^* dA + L_3^* dJ^* + L_4^* dJ)(L_1 dA + L_2 dA^* + L_3 dJ + L_4 dJ^*)U$$

$$= U^*(-2Pdt + (L_1^* + L_2)dA^* + (L_2^* + L_1)dA + (L_3^* + L_4)dJ^*$$

$$+(L_3 + L_4^*)dJ + (L_2^* L_2 + L_4^* L_4)dt + L_2^* L_4(-1)^\Lambda dt + L_4^* L_2(-1)^\Lambda dt)U$$

This gives

$$P = (1/2)(L_2^* L_2 + L_4^* L_4 + (-1)^\Lambda (L_2^* L_4 + L_4^* L_2)),$$

$$L_1^* + L_2 = 0, L_3^* + L_4 = 0$$

as the condition for unitary evolution.

$$j_t(X) = U(t)^* X U(t)$$

$$dj_t(X) = dU(t)^* X U(t) + U(t)^* X dU(t) + dU(t)^* X dU(t)$$

$$= U^*([(iH-P)dt + L_1^* dA^* + L_2^* dA + L_3^* dJ^* + L_4^* dJ]X + X[-(iH+P)dt + L_1 dA$$

$$+L_2 dA^* + L_3 dJ + L_4 dJ^*])U$$

$$+U^*(L_1^* dA^* + L_2^* dA + L_3^* dJ^* + L_4^* dJ)X(L_1 dA + L_2 dA^* + L_3 dJ + L_4 dJ^*)U$$

$$= U^*(i[H,X]dt - (PX + XP)dt + (L_1^* X + X L_2)dA^* + (L_2^* X + X L_1)dA$$

$$+$$

$$+(L_3^* X + X L_4)dJ^* + (X L_3 + L_4^* X)dJ + (L_2^* X L_2 + L_4^* X L_4)dt$$

$$+L_2^* X L_4(-1)^\Lambda dt + L_4^* X L_2(-1)^\Lambda dt)U$$

These equations can be cast in the form

$$dj_t(X) = j_t(\theta_0(X) + \theta_1(X)(-1)^\Lambda)dt + j_t(\theta_2(X))dA + j_3(\theta_3(X))dA^*$$

Assume that the input measurement process is

$$dY_i(t) = c_1 dA(t) + \bar{c}_1 dA(t)^* + c_2\theta(t)dJ(t) + \bar{c}_2\theta(t)^* dJ(t)^*$$

where the $\theta(t)'s$ are anticommuting Fermionic parameters. These parameters are also assumed to anticommute wtih the processes $A(s), A(s)^*, s \leq t$. We shall check whether $Y_i(t), t \geq 0$ forms an Abelian family of operators. For this we require first that $[\theta(t)dJ(t), \theta(s)dJ(s)] = 0$ for all s, t. Now $dJ(t)$ anticommutes with $dJ(s)$ and $\theta(t)$ anticommutes with $\theta(s)$. If we assume that $\theta(t)$ anticommutes with $dJ(s)$, then this commutator will be zero. Of course it will also be zero if $\theta(t)$ commutes with $dJ(s)$ but we choose the latter. For Y_i to form an Abelian family, we also require that $[dA(t), \theta(s)dJ(s)] = 0$ for all s, t. Now, for $t \leq s$, $dA(t)$ anticommutes with $dJ(s)$ so we require that $\theta(s)$ anticommute with $dA(t)$ for $t < s$. For $t > s$, $dA(t)$ commutes with $dJ(s)$, so we require that $\theta(s)$ commutes with $dA(t)$ for $t > s$. In summary, for $Y_i(.)$ to form an Abelian family, we require that $\theta(s)$ and $\theta(s)^*$ commute with $dA(t), dA(t)^*$ for $t \geq s$, anticommute with the same for $t < s$ and anticommute with $dJ(t), dJ(t)^*$ for all t.

Then we have

$$\theta(t)dJ(t)A(s) = -\theta(t)A(s)dJ(t) = A(s)\theta(t)dJ(t), s \leq t$$

The output measurement process is

$$Y_o(t) = U(t)^*Y_i(t)U(t)$$

and for it to be non demolition, we require

$$d_T(U(T)^*Y_i(t)U(T)) = 0, T \geq t$$

ie,

$$dU(T)^*Y_i(t)U(T) + U(T)^*Y_i(t)dU(T) + dU(T)^*Y_i(t)dU(T) = 0, T > t$$

or equivalently,

$$U(T)^*(-(iH+P(T))dT+L_1^*dA^*(T)+L_2^*dA(T)+L_3*dJ^*(T)+L_4^*dJ(T))Y_i(t)U(T)$$

$$+U(T)^*Y_i(t)(-(iH+P(T))dT+L_1dA(T)+L_2dA(T)^*+L_3dJ(T)+L_4dJ(T)^*)U(T)$$

$$+U(T)^*(L_2^*dA(T) + L_4^*dJ(T))Y_i(t)(L_2dA(T)^* + L_4dJ(T)^*U(T) = 0, T \geq t$$

For this, we first of all require that $P(T)$ should commute with $Y_i(t)$ and this amounts to $(-1)^{\Lambda(T)}$ to commute with $Y_i(t)$ for $T \geq t$. But actually the component $A(t)$ in $Y_i(t)$ anticommutes with $(-1)^{\Lambda(T)}$. So to rectify this, we incorporate the anticommuting Fermionic parameters $\theta(t)$ in $dJ(t)$ and likewise $\theta(t)^*$ in $dJ(t)^*$ in the HP qsde (This is the second reason for incorporating the Fermionic parameters, earlier to achieve Abelian input measurements we had incorporated Fermionic parameters in the measurement model, now we also incorporate it in the HP qsde to ensure the non-demolition property of the output measurement process). Then the HP equation then becomes

$$dU = (-(iH + P)dt + L_1dA + L_2dA^* + L_3\theta.dJ + L_4\theta^*.dJ^*)U$$

where $\theta = \theta(t)$ etc. Then, the last condition for unitary of $U(t)$ becomes

$$P = (1/2)(L_2^*L_2 + |\theta(t)|^2L_4^*L_4) + (-1)^{\Lambda(t)}(\theta(t)^*L_2^*L_4 + \theta(t)L_4^*L_2))$$

since now

$$\theta(t).dJ(t).\theta(t)^*.dJ(t)^* = -\theta(t)\theta(t)^*.dJ(t).dJ(t)^*$$

$$= -\theta(t)\theta(t)^*dt = \theta(t)^*\theta(t)dt = |\theta(t)|^2dt$$

Note that we have used

$$\theta(t)dJ(t).dA(t)^* = \theta(t)(-1)^{\Lambda(t)}dt,$$

$$dA(t)\theta(t)dJ(t)^* = \theta(t)dA(t)dJ(t)^* = \theta(t)(-1)^{\Lambda(t)}dt$$

since $\theta(t)$ commutes with $dA(t)$.

The evolution equation for $j_t(X)$ will accordingly get modified to

$$dj_t(X) =$$

$$= U^*([(iH-P)dt+L_1^*dA^*+L_2^*dA+\theta^*L_3^*dJ^*+\theta L_4^*dJ]X+X[-(iH+P)dt$$
$$+L_1dA+L_2dA^*+\theta L_3dJ+\theta^*L_4dJ^*])U$$
$$+U^*(L_1^*dA^*+L_2^*dA+\theta^*L_3^*dJ^*+\theta L_4^*dJ)X(L_1dA+L_2dA^*+\theta L_3dJ+\theta^*L_4dJ^*)U$$
$$= U^*(i[H,X]dt-(PX+XP)dt+(L_1^*X+XL_2)dA^*$$
$$+(L_2^*X+XL_1)dA+\theta^*(L_3^*X+XL_4)dJ^*+\theta(XL_3+L_4^*X)dJ+$$
$$(L_2^*XL_2+|\theta|^2L_4^*XL_4)dt+$$
$$+\theta^*L_2^*XL_4(-1)^\Lambda dt+\theta L_4^*XL_2(-1)^\Lambda dt)U$$

The non-demolition condition after incorporation of the Fermionic parameters becomes

$$U(T)^*(-(iH+P(T))dT+L_1^*dA^*(T)+L_2^*dA(T)+\theta(T)^*L_3*dJ^*(T)+\theta(T)L_4^*dJ(T))Y_i(t)U(T)$$
$$+U(T)^*Y_i(t)(-(iH+P(T))dT+L_1dA(T)+L_2dA(T)^*+\theta(T)L_3dJ(T)+\theta(T)^*L_4dJ(T)^*)U(T)$$
$$+U(T)^*(L_2^*dA(T)+\theta(T)L_4^*dJ(T))Y_i(t)(L_2dA(T)^*+\theta(T)^*L_4dJ(T)^*U(T)=0, T\geq t$$

and the second condition for non-demolition that we require is that $Y_i(t)$ should commute with $\theta(T)^*dJ(T)^*$ and with $\theta(T)dJ(T)$ which is true since $\theta(T)(-1)^{\Lambda(T)}$ commutes with $dA(t), dA(t)^*, \theta(t)(-1)^{\Lambda(t)}dA(t)$ and with $\theta(t)^*(-1)^{\Lambda(t)}dA(t)^*$ for $T \geq t$.

Note: $\theta(T)dJ(T)$ commutes with $dA(t)$ for $T > t$ since $\theta(T)$ anticommutes with $dA(t)$ and $dJ(T)$ also anticommutes with $dA(t)$. $\theta(T)dJ(T)$ commutes with $\theta(t)dJ(t)$ since $\theta(T)$ and $dJ(T)$ both anticommute with each of $\theta(t)$ and $dJ(t)$. It should be noted that $\theta(t)$ anticommutes with $dJ(T) = (-1)^{\Lambda(T)}dA(T)$ and commutes with $dA(T)$ for $T \geq t$ implies that $\theta(t)$ anticommutes with $(-1)^{\Lambda(T)}$ for $T \geq t$. Also, $\theta(T)$ anticommutes with $dJ(t) = (-1)^{\Lambda(t)}dA(t)$ and with $dA(t)$ for $T > t$ implies that $\theta(T)$ commutes with $(-1)^{\Lambda(t)}$ for $T > t$. Further, $dA(T)$ commutes with $dJ(t) = (-1)^{\Lambda(t)}dA(t)$ for $T > t$ and also commutes with $\theta(t)dJ(t)$ for $T > t$ implies that $dA(T)$ commutes with $\theta(t)$ for $T > t$. Thus the entire picture is self-consistent.

The Boson-Fermion quantum filter: Let

$$\eta_o(t) = \sigma(Y_o(s) : s \leq t), Y_o(t) = U(t)^*Y_i(t)U(t) = U(T)^*Y_i(t)U(T), T \geq t$$

Assume that the filter equations for

$$\pi_t(X) = \mathbb{E}(j_t * (X)|\eta_o(t))$$

are given by

$$d\pi_t(X) = F_t(X)dt + G_t(X)dY_o(t)$$

wher $F_t(X), G_t(X)$ are $\eta_o(t)$-measurable. Note that

$$dY_o(t) = d(U(t)^*Y_i(t)U(t))$$
$$= U(t)^*dY_i(t)U(t)+dU(t)^*Y_i(t)U(t)+U(t)^*Y_i(t)dU(t)+dU(t)^*Y_i(t)dU(t)$$

Note that we are assuming that each of $\theta(t), \theta(t)^*$ anticommutes with each of $dA(s), dA(s)^*$ for $s < t$ and commutes with $dA(s), dA(s)^*$ for $s \geq t$. Now

$$U(t)^*dY_i(t)U(t) = U(t)^*(c_1 dA(t) + \bar{c}_1 dA(t)^* + c_2\theta(t)dJ(t) + \bar{c}_2\theta(t)^*dJ(t)^*)U(t)$$

Now, $dA(t), dA(t)^*$ both commute with $dA(s), dA(s)^*$ and also with $\theta(s)dJ(s), \theta(s)^*dJ(s)^*$ for $s < t$ and therefore, $U(t)^*$ commutes with $dA(t), dA(t)^*$. Likewise, $\theta(t)dJ(t), \theta(t)^*dJ(t)^*$ both commute with $dA(s), dA(s)^*$ and also with $\theta(s)dJ(s), \theta(s)^*dJ(s)^*$ for $s < t$. Thus, $dY_i(t)$ commutes with each of $\{dA(s), dA(s)^*, \theta(s)dJ(s), \theta(s)^*dJ(s)^*\}$ for $t > s$ and hence $U(t)$ commutes with $dY_i(t)$. This implies

$$U(t)^*dY_i(t)U(t) = dY_i(t).U(t)^*U(t) = dY_i(t)$$

by the unitarity of $U(t)$.

Note: It is imperative that $\theta(s), \theta(s)^*$ commute with $dA(t), dA(t)^*$ for $t \geq s$ because of the following reason: For $T > t$,

$$d_T U(T)^*Y_i(t)U(T) = dU(T)^*Y_i(t)U(T)+U(T)^*Y_i(t)dU(T)+dU(T)^*Y_i(t)dU(T)$$

For non-demolition, we require this to be zero. Since $U(T)$ is unitary, this condition boils down to requiring that $Y_i(t)$ and hence $dA(s), dA(s)^*, \theta(s)dJ(s), \theta(s)^*dJ(s)^*$ all commute with $dA(T), dA(T)^*, \theta(T)dJ(T), \theta(T)^*dJ(T)^*$ for all $s < T$. Then for example since $dA(s)$ is required to commute with $\theta(T)dJ(T)$ while $dA(s)$ anticommutes with $dJ(T)$, it follows that $dA(s)$ must anticommute with $\theta(T)$ for $T > s$. Likewise, since $\theta(s)dJ(s)$ is required to commute with $dA(T)$ for $T > s$ and since $dJ(s)$ commutes with $dA(T)$, we require that $\theta(s)$ commute with $dA(T)$ for $T > s$.

We also note that since $\theta(s)dJ(s)$ commutes with $dA(t)$ for $t > s$ and since $dJ(s)$ commutes with $dA(t)$ for $t > s$, it follows that $\theta(s)$ must commute with $dA(t)$ for $t > s$. Since $\theta(s)$ is assumed to anticommute with $dJ(t) = (-1)^{\Lambda(t)}dA(t)$ for all t, and $\theta(s)$ commutes with $dA(t)$ for $t > s$, it follows that $\theta(s)$ must necessarily anticommute with $(-1)^{\Lambda(t)}$ for $t > s$. The same holds for $\theta(s)^*$. Further, $\theta(t)$ is assumed to anticommute with $dJ(s) = (-1)^{\Lambda(s)}dA(s)$ for all s and in particular for $t > s$ and $\theta(t)$ anticommutes with $dA(s)$ for $t > s$, it follows that $\theta(t)$ must necessarily commute with $(-1)^{\Lambda(s)}$ for $t > s$. Note that $\theta(t)$ must anticommute with $dA(s)$ for $t > s$ because we require that $\theta(t)dJ(t)$ commute with $dA(s)$ for $t > s$ while $dJ(t)$ anticommutes with $dA(s)$ for $t > s$.

Remark: Suppose that $\theta(t)$ is assumed to commute with $dJ(s)$ for all t, s. Then, $\theta(t)dJ(t)$ would anticommute with $dJ(s)$ for $t > s$ since $dJ(t)$ anticommutes with $dJ(s)$ for all t, s. But then,

$$\theta(t)dJ(t).\theta(s).dJ(s) = \theta(t)\theta(s)dJ(t)dJ(s) =$$
$$\theta(s)\theta(t)dJ(s)dJ(t) = \theta(s)dJ(s)\theta(t)dJ(t)$$

as required by the non-demolition condition. However, the non-demolition condition also requires that $\theta(t)dJ(t)$ also commute with $dA(s)$ for all t, s. This would imply that $\theta(t)$ anticommute with $dA(s)$ for $t > s$ (since $dJ(t)$ anticommutes with $dA(s)$ for $t > s$) and commute with $dA(s)$ for $s > t$ (since $dJ(t)$ commutes with $dA(s)$ for $s > t$). Likewise, $\theta(t)$ would anticommute with $dA(s)^*$ for $t > s$ and commute with $dA(s)^*$ for $s > t$. Then, $\theta(t)$ would commute with $\Lambda(s) = \int_0^s dA(u)^* dA(u)/du$ and hence with $(-1)^{\Lambda(s)}$ for $t > s$. We could thus develop a Boson-Fermion filtering theory based on this assumption also. This could be seen in another way also as follows. Since $\theta(t)$ is assumed to commute with $dJ(s) = (-1)^{\Lambda(s)} dA(s)$ for all t, s and since we've seen that $\theta(t)$ commutes with $dA(s)$ for $s > t$, it follows that $\theta(t)$ must commute with $(-1)^{\Lambda(s)}$ for $s > t$ (since $\theta(t)$ is assumed to commute with $dJ(s)$). On the other hand, since $\theta(t)$ anticommutes with $dA(s)$ for $t > s$, it must follow that $\theta(t)$ anticommutes with $(-1)^{\Lambda(s)}$ for $t > s$. There is thus a contradiction involved here.

This contradiction is resolved if we assume that $\theta(t)$ anticommutes (rather than commutes) with $dJ(s)$ for all s. For we then get that

$$\theta(t)dJ(t).\theta(s)dJ(s) = -\theta(t)\theta(s)dJ(t)dJ(s) = -\theta(s)\theta(t)dJ(s)dJ(t)$$

$$= \theta(s)dJ(s)\theta(t)dJ(t)$$

for all t, s as required by the non-demolition property. Further, the non-demolition property requires that $\theta(t)dJ(t)$ commute with $dA(s)$ for all t, s and therefore that $\theta(t)$ anticommute with $dA(s)$ for $t > s$ and commute with $dA(s)$ for $s > t$. Likewise, $\theta(t)$ should then anticommute with $dA(s)^*$ for $t > s$ and commute with $dA(s)^*$ for $s > t$. This then implies that $\theta(t)$ commutes with $\Lambda(s) = \int_0^s dA(u)^* dA(u)/du$ for $t > s$. On the other hand, since $\theta(t)$ is assumed to anticommute with $dJ(s) = (-1)^{\Lambda(s)} dA(s)$ for all t, s while $\theta(t)$ anticommutes with $dA(s)$ for $t > s$ and commutes with $dA(s)$ for $s > t$, it would follow that $\theta(t)$ anticommutes with $(-1)^{\Lambda(s)}$ for $s > t$ and commutes with the same for $t > s$. The earlier contradiction is therefore resolved here.

The quantum Boson-Fermion filter:

$$d\pi_t(X) = F_t(X)dt + G_t(X)dY_o(t)$$

$$dC(t) = f(t)C(t)dY_o(t), C(0) = 1$$

$$\mathbb{E}[(j_t(X) - \pi_t(X))C(t)] = 0$$

$$dj_t(X) = j_t(\theta_0(X) + \theta_1(X)(-1)^{\Lambda(t)})dt$$

$$+ j_t(\theta_2(X))dA(t) + j_t(\theta_3(X))dA(t)^*$$

$$+ j_t(\theta_4(X))\theta(t)dJ(t) + j_t(\theta_5(X))\theta(t)^* dJ(t)^*$$

where $\theta_0(X)$ contains terms involving $|\theta(t)|^2$ while $\theta_1(X)$ contains terms linear in $\theta(t), \theta(t)^*$. Now

$$\mathbb{E}[(dj_t(X) - d\pi_t(X))C(t)] + \mathbb{E}[(j_t(X) - \pi_t(X))dC(t)] + \mathbb{E}[(dj_t(X) - d\pi_t(X))dC(t)] =$$

From this equation and the arbitrariness of the complex valued function $f(t)$, we infer that

$$\mathbb{E}[(dj_t(X) - d\pi_t(X))C(t)] = 0,$$
$$\mathbb{E}[(j_t(X) - \pi_t(X))C(t)dY_o(t)] + \mathbb{E}[(dj_t(X) - d\pi_t(X))C(t)dY_o(t)] = 0$$

Therefore

$$\mathbb{E}[(dj_t(X) - d\pi_t(X))|\eta_o(t)] = 0,$$
$$\mathbb{E}[(j_t(X) - \pi_t(X))dY_o(t)|\eta_o(t)] + \mathbb{E}[(dj_t(X) - d\pi_t(X))dY_o(t)|\eta_o(t)] = 0$$

We define in addition

$$\nu_t(X) = \mathbb{E}(j_t(X)(-1)^{\Lambda(t)}|\eta_o(t))$$

Then,

$$d(j_t(X).(-1)^{\Lambda(t)}) = dj_t(X).(-1)^{\Lambda(t)} - 2j_t(X)(-1)^{\Lambda(t)}d\Lambda(t)$$
$$-2dj_t(X)d\Lambda(t).(-1)^{\Lambda(t)}$$
$$= [j_t(\theta_0(X) + \theta_1(X)(-1)^{\Lambda(t)})dt + j_t(\theta_2(X))dA(t) + j_t(\theta_3(X))dA(t)^*$$
$$+j_t(\theta_4(X))\theta(t)dJ(t) + j_t(\theta_5(X))\theta(t)^*dJ(t)^*](-1)^{\Lambda(t)}$$
$$-2j_t(X)(-1)^{\Lambda(t)}d\Lambda(t)$$
$$-2[j_t(\theta_0(X) + \theta_1(X)(-1)^{\Lambda(t)})dt$$
$$+j_t(\theta_2(X))dA(t) + j_t(\theta_3(X))dA(t)^*$$
$$+j_t(\theta_4(X))\theta(t)dJ(t) + j_t(\theta_5(X))\theta(t)^*dJ(t)^*]d\Lambda(t)(-1)^{\Lambda(t)}$$

We then observe that by quantum Ito's formula,

$$dJ(t)d\Lambda(t) = (-1)^{\Lambda(t)}dA(t) = dJ(t), dA(t)d\Lambda(t) = dA(t),$$
$$dJ(t)^*d\Lambda(t) = 0, dA(t)^*d\Lambda(t) = 0,$$
$$dJ(t)(-1)^{\Lambda(t)} = dA(t), dJ(t)^*(-1)^{\Lambda(t)} = dA(t)^*$$
$$dJ(t) = (-1)^{\Lambda(t)}dA(t), dJ(t)^* = (-1)^{\Lambda(t)}dA(t)^*$$

and thus the above simplifies to

$$d(j_t(X).(-1)^{\Lambda(t)}) =$$
$$[j_t(\theta_1(X))dJ(t) + j_t(\theta_2(X))dJ(t)^*$$
$$+j_t(\theta_2(X))\theta(t)dA(t) + j_t(\theta_3(X))\theta(t)^*dA(t)^*]$$
$$-2j_t(X)(-1)^{\Lambda(t)}d\Lambda(t)$$
$$-2[j_t(\theta_2(X))dJ(t) + j_t(\theta_4(X))\theta(t)dA(t)]$$

Observe that

$$dj_t(X) = j_t(\theta_0(X))dt + j_t(\theta_1(X))dA(t) + j_t(\theta_2(X))dA(t)^*$$

where

$$\theta_0(X) = \theta_{00}(X) + \theta_{01}(X)(-1)^{\Lambda(t)},$$
$$\theta_1(X) = \theta_{10}(X) + \theta_{11}(X)(-1)^{\Lambda(t)}$$
$$\theta_2(X) = \theta_{20}(X) + \theta_{21}(X)(-1)^{\Lambda(t)}$$

where $\theta_{ab}(X)$ are system space operators containing the Fermionic parameters $\theta(t), \theta(t)^*$. We can also write

$$dY_i(t) = (c_1 + c_2\theta(t)(-1)^{\Lambda(t)})dA(t) + (\bar{c}_1 + \bar{c}_2\theta(t)^*(-1)^{\Lambda(t)})dA(t)^*$$
$$= \lambda(t)dA(t) + \lambda(t)^*dA(t)^*$$

with

$$\lambda(t) = c_1 + c_2\theta(t)(-1)^{\Lambda(t)}$$

Then

$$dY_o(t) = j_t(M)dt + \lambda_1.dA + \lambda_2 dA^*,$$
$$M = M_0 + (\theta M_{10} + \theta^* M_{11})(-1)^{\Lambda} = M_0 + M_1(-1)^{\Lambda}$$
$$\lambda_1 = \lambda_{10} + \tilde{\lambda}_{11}\theta(-1)^{\Lambda}, = \lambda_{10} + \lambda_{11}(-1)^{\Lambda(t)}$$
$$\lambda_2 = \lambda_{20} + \tilde{\lambda}_{21}\theta^*(-1)^{\Lambda} = \lambda_{20} + \lambda_{21}(-1)^{\Lambda(t)}$$

Note that the Boson-Fermion HP qsde can be expressed as

$$dU(t) = (-(iH + P)dt + L_1 dA + L_2 dA^*)U(t)$$

where

$$L_1 = L_{10} + L_{11}\theta(-1)^{\Lambda},$$
$$L_2 = L_{20} + L_{21}\theta^*(-1)^{\Lambda}$$
$$P = P_0 + P_1(-1)^{\Lambda}$$

where

$$P_0 = P_{00} + \theta P_{01}\theta + P_{02}\theta^*$$
$$P_1 = P_{10} + P_{11}|\theta|^2$$

with L_{ab}, P_{ab} all being system space operators not involving the Fermionic parameters θ, θ^*. We define for a system space operator X, the following conditional expectations:

$$\pi_t(X) = \mathbb{E}[j_t(X)|\eta_o(t)], \nu_t(X) = \mathbb{E}[j_t(X)(-1)^{\Lambda(t)}|\eta_o(t)],$$
$$\rho_t(X) = \mathbb{E}[j_t(X.(-1)^{\Lambda(t)})|\eta_o(t)],$$
$$\sigma_t(X) = \mathbb{E}[j_t(X.(-1)^{\Lambda(t)})(-1)^{\Lambda(t)}|\eta_o(t)]$$

and we derive differential equations for these.

$$\mathbb{E}[dj_t(X)|\eta_o(t)] = \mathbb{E}[j_t(\theta_0(X))dt + j_t(\theta_1(X))dA(t) + j_t(\theta_2(X))dA(t)^*|\eta_o(t)]$$

$$= \mathbb{E}[j_t(\theta_{00}(X) + \theta_{01}(X)(-1)^{\Lambda(t)})|\eta_o(t)]dt$$
$$+u(t)dt\mathbb{E}[j_t(\theta_{10}(X) + \theta_{11}(X)(-1)^{\Lambda(t)})|\eta_o(t)]$$
$$+\bar{u}(t)dt\mathbb{E}[j_t(\theta_{20}(X) + \theta_{21}(X)(-1)^{\Lambda(t)})|\eta_o(t)]$$
$$= dt[\pi_t(\theta_{00}(X)) + \rho_t(\theta_{01}(X))] +$$
$$u(t)dt[\pi_t(\theta_{10}(X)) + \rho_t(\theta_{11}(X))] +$$
$$\bar{u}(t)dt[\pi_t(\theta_{20}(X)) + \rho_t(\theta_{21}(X))]$$
$$\mathbb{E}[j_t(X)dY_o(t)|\eta_o(t)] =$$
$$\mathbb{E}[j_t(X)(j_t(M)dt + \lambda_1.dA + \lambda_2 dA^*)|\eta_o(t)]$$
$$= dt\mathbb{E}[j_t(XM)|\eta_o(t)] + u(t)dt\mathbb{E}[j_t(X)\lambda_1|\eta_o(t)]$$
$$+\bar{u}(t)dt\mathbb{E}[j_t(X)\lambda_2|\eta_o(t)]$$

Now,

$$\mathbb{E}[j_t(XM)|\eta_o(t)] = \mathbb{E}[j_t(X(M_0 + M_1(-1)^{\Lambda(t)}))|\eta_o(t)]$$
$$= \pi_t(XM_0) + \rho_t(XM_1)$$
$$\mathbb{E}[j_t(X)\lambda_1|\eta_o(t)] = \pi_t(X)\lambda_{10} + \nu_t(X)\lambda_{11},$$
$$\mathbb{E}[j_t(X)\lambda_2|\eta_o(t)] = \pi_t(X)\lambda_{20} + \nu_t(X)\lambda_{21},$$

Thus,

$$\mathbb{E}[j_t(X)dY_o(t)|\eta_o(t)] =$$
$$[\pi_t(XM_0) + \rho_t(XM_1) + \pi_t(X)\lambda_{10} + \nu_t(X)\lambda_{11} + \pi_t(X)\lambda_{20} + \nu_t(X)\lambda_{21}]dt$$

We write our filtering equations as

$$d\pi_t(X) = F_{1t}(X)dt + G_{1t}(X)dY_o(t),$$
$$d\nu_t(X) = F_{2t}(X)dt + G_{2t}(X)dY_o(t),$$
$$d\rho_t(X) = F_{3t}(X)dt + G_{3t}(X)dY_o(t),$$
$$d\sigma_t(X) = F_{4t}(X)dt + G_{4t}(X)dY_o(t)$$

Note that all the quantities $F_{kt}(X), G_{kt}(X), k = 1, 2, 3, 4$ are in $\eta_o(t)$ and are therefore commutative. The orthogonality principle in estimation theory states that

$$\mathbb{E}[(j_t(X) - \pi_t(X))C(t)] = 0,$$
$$\mathbb{E}[(j_t(X)(-1)^{\Lambda(t)} - \nu_t(X))C(t)] = 0,$$
$$\mathbb{E}[(j_t(X.(-1)^{\Lambda(t)}) - \rho_t(X))C(t)] = 0,$$
$$\mathbb{E}[(j_t(X.(-1)^{\Lambda(t)}).(-1)^{\Lambda(t)} - \sigma_t(X))C(t)] = 0$$

where $C(t)$ is any $\eta_o(t)$ measurable observable. In particular, this is true for

$$dC(t) = f(t)C(t)dY_o(t), t \geq 0, C(0) = 1$$

By this orthogonality principle in estimation theory, we observe that on taking differentials in time, using $d(\xi.\eta) = d\xi.\eta + \xi.d\eta + d\xi.d\eta$ and choosing f appropriately that

$$\mathbb{E}[dj_t(X) - d\pi_t(X)|\eta_o(t)] = 0,$$

$$\mathbb{E}[(j_t(X) - \pi_t(X))dY_o(t)|\eta_o(t)] + \mathbb{E}[dj_t(X) - d\pi_t(X))dY_o(t)|\eta_o(t)] = 0$$

$$\mathbb{E}[dj_t(X.(-1)^{\Lambda(t)}) - d\rho_t(X)|\eta_o(t)] = 0,$$

$$\mathbb{E}[(j_t(X(-1)^{\Lambda(t)}) - \rho_t(X))dY_o(t)|\eta_o(t)] + \mathbb{E}[dj_t(X.(-1)^{\Lambda(t)}) - d\rho_t(X))dY_o(t)|\eta_o(t)] = 0$$

$$\mathbb{E}[d(j_t(X)(-1)^{\Lambda(t)}) - d\nu_t(X)|\eta_o(t)] = 0,$$

$$\mathbb{E}[(j_t(X)(-1)^{\Lambda(t)}) - \nu_t(X))dY_o(t)|\eta_o(t)] + \mathbb{E}[d(j_t(X)(-1)^{\Lambda(t)}) - d\nu_t(X))dY_o(t)|\eta_o(t)] = 0$$

$$\mathbb{E}[d(j_t(X.(-1)^{\Lambda(t)})(-1)^{\Lambda(t)}) - d\nu_t(X)|\eta_o(t)] = 0,$$

$$\mathbb{E}[(j_t(X.(-1)^{\Lambda(t)})(-1)^{\Lambda(t)} - \sigma_t(X))dY_o(t)|\eta_o(t)] + \\ \mathbb{E}[d(j_t(X(-1)^{\Lambda(t)})(-1)^{\Lambda(t)}) - d\sigma_t(X))dY_o(t)|\eta_o(t)] = 0$$

Now

$$\begin{gathered}\mathbb{E}[dj_t(X)|\eta_o(t)] = \\ dt[\pi_t(\theta_{00}(X)) + \rho_t(\theta_{01}(X)) + \\ u(t)(\pi_t(\theta_{10}(X)) + \rho_t(\theta_{11}(X))) + \\ \bar{u}(t)(\pi_t(\theta_{20}(X)) + \rho_t(\theta_{21}(X)))] \\ \mathbb{E}[dj_t(X).dY_o(t)|\eta_o(t)] = \\ \mathbb{E}[j_t(\theta_1(X))dA(t).\lambda_2.dA^*(t)|\eta_o(t)] \\ = dt.\mathbb{E}[j_t(\theta_{10}(X) + \theta_{11}(X)(-1)^{\Lambda(t)}).(\lambda_{20} + \lambda_{21}(-1)^{\Lambda(t)})|\eta_o(t)] \\ = dt.[\pi_t(\theta_{10}(X))\lambda_{20} + \nu_t(\theta_{10}(X))\lambda_{21} \\ +\rho_t(\theta_{11}(X))\lambda_{20} + \sigma_t(\theta_{11}(X))\lambda_{21}]\end{gathered}$$

Next,

$$\begin{gathered}d(j_t(X)(-1)^{\Lambda(t)}) = \\ [j_t(\theta_0(X))dt + j_t(\theta_1(X))dA(t) + j_t(\theta_2(X))dA(t)^*](-1)^{\Lambda} \\ -2j_t(X)(-1)^{\Lambda(t)}d\Lambda(t) \\ -2j_t(\theta_1(X))(-1)^{\Lambda(t)}dA(t)d\Lambda(t)\end{gathered}$$

Thus, using $dA.d\Lambda = dA$, we get

$$\mathbb{E}[d(j_t(X)(-1)^{\Lambda(t)})|\eta_o(t)] =$$

$$dt[\nu_t(\theta_{00}(X)) + \sigma_t(\theta_{01}(X)) + u(t)(\nu_t(\theta_{10}(X)) + \sigma_t(\theta_{11}(X))) + \bar{u}(t)(\nu_t(\theta_{20}(X)) + \sigma_t(\theta_{21}(X)))] \\ -2dt|u(t)|^2\nu_t(X) - 2u(t)dt(\nu_t(\theta_{10}(X)) + \sigma_t(\theta_{11}(X)))$$

Similarly by use of the quantum Ito formula, we can calculate

$$d(j_t(X)(-1)^{\Lambda(t)}).dY_o(t), d(j_t(X.(-1)^{\Lambda(t)})dY_o(t), d(j_t(X.(-1)^{\Lambda(t)}).(-1)^{\Lambda(t)}).dY_o$$

in terms of $dt, dA, dA^*, d\Lambda$ and hence derive equations for $F_{kt}(.), G_{kt}(.)$. We leave this as an exercise reader. Note that this analysis immediately leads to stochastic coupled differential equations for $\pi_t, \rho_t, \nu_t, \sigma_t$. These equations have the following form:

$$d\pi_t(X) = F_1(\pi_t(\alpha_1(X)), \nu_t(\alpha_2(X)), \rho_t(\alpha_3(X)), \sigma_t(\alpha_4(X))dt$$

$$+F_2(\pi_t(\beta_1(X)), \nu_t(\beta_2(X)), \rho_t(\beta_3(X)), \sigma_t(\beta_4(X))dY_o(t)$$

and likewise for ν_t, ρ_t, σ_t where the maps $\alpha_k, \beta_k, k = 1, 2, 3, 4$ depend upon $u(t)$ which parametrizes the coherent state. Thus, we obtain concrete formulas for the quantum filter.

$$dj_t(X.(-1)^{\Lambda(t)}) = d(U(t)^* X.(-1)^{\Lambda(t)}.U(t))$$

$$= dU(t)^*.X.(-1)^{\Lambda(t)}.U(t) + U(t)^* X.(-1)^{\Lambda(t)}.dU(t)$$

$$+dU(t)^* X.(-1)^{\Lambda(t)}.dU(t) - 2U(t)^*.X(-1)^{\Lambda(t)}U(t)d\Lambda(t)$$

$$+dU(t)^* d\Lambda(t).X.(-1)^{\Lambda(t)}U(t) + U(t)^* X.(-1)^{\Lambda(t)}d\Lambda(t).dU(t)$$

It is clear that this can be expressed as

$$j_t(\phi_{00}(X) + \phi_{01}(X)(-1)^{\Lambda(t)}))dt+$$

$$j_t(\phi_{10}(X) + \phi_{11}(X)(-1)^{\Lambda(t)})dA(t)+$$

$$j_t(\phi_{20}(X) + \phi_{21}(X)(-1)^{\Lambda(t)})dA(t)^*+$$

$$-2j_t(X.(-1)^{\Lambda(t)})d\Lambda(t)$$

and likewise,

$$d(j_t(X)(-1)^{\Lambda(t)}) = j_t(\theta_{00}(X) + \theta_{01}(X)(-1)^{\Lambda(t)})(-1)^{\Lambda(t)}dt$$

$$+j_t(\theta_{10}(X) + \theta_{11}(X)(-1)^{\Lambda(t)})(-1)^{\Lambda(t)}dA(t)$$

$$+j_t(\theta_{20}(X) + \theta_{21}(X)(-1)^{\Lambda(t)})(-1)^{\Lambda(t)}dA(t)^*$$

$$-2j_t(\theta_{10}(X) + \theta_{11}(X)(-1)^{\Lambda(t)})(-1)^{\Lambda(t)}dA(t)$$

$$-2j_t(X)(-1)^{\Lambda(t)}d\Lambda(t)$$

Recall that

$$dj_t(X) = j_t(\theta_{00}(X) + \theta_{01}(X)(-1)^{\Lambda(t)})dt+$$

$$j_t(\theta_{10}(X) + \theta_{11}(X)(-1)^{\Lambda(t)})dA(t)+$$

$$j_t(\theta_{20}(X) + \theta_{11}(X)(-1)^{\Lambda(t)})dA(t)^*$$

Thus, we get in addition

$$d(j_t(X.(-1)^{\Lambda(t)})(-1)^{\Lambda(t)})$$

1.6 Quantum field theory is a low energy limit of string field theory

Example: Consider a Bosonic string interacting with a background metric field $g_{\mu\nu}(X)$ and also with a background antisymmetric gauge field $B_{\mu\nu}(X)$. The action functional is given by

$$S[X|g,B] = \int (1/2)g_{\mu\nu}(X)\eta^{\alpha\beta}X^{\mu}_{,\alpha}X^{\nu}_{,\beta}d^2\sigma$$

$$+\int B_{\mu\nu}(X)\epsilon^{\alpha\beta}X^{\mu}_{,\alpha}X^{\nu}_{,\beta}d^2\sigma$$

Here, $X = ((X^{\mu}(\tau,\sigma)))_{\mu=0}^{D-1}$ is the string field. We write it as

$$X^{\mu}(\tau,\sigma) = x^{\mu} + \delta X^{\mu}(\tau,\sigma)$$

The string propagator is in flat space-time and in the absence of interactions is given by

$$< T[\delta X^{\mu}(\tau,\sigma).\delta X^{\nu}(\tau,\sigma)] >= D^{\mu\nu}(\tau,\sigma|\tau',\sigma') = K.\eta^{\mu\nu}ln(\sqrt{(\tau-\tau')^2+(\sigma-\sigma')^2}$$

Using this, we can calculate approximately the classical action for fields g, B.

$$S_0[g,B] =< S[X|g,B] >$$

We can also calculate the quantum effective action using the Feynman path integral

$$exp(iS_1[g,B]/h) = \int exp(iS_0[X|g,B]/h)DX$$

and expand $S_1[g,B]$ in powers of h and then hope that the coefficients of different powers of h will yield quantum string theoretic corrections to the classical actions for g, B.

Note: Let

$$S[X] = \int B_{\mu\nu}(X)dX^{\mu}\wedge dX^{nu} = \int B_{\mu\nu}(X)\epsilon^{\alpha\beta}X^{\mu}_{,\alpha}X^{\nu}_{,\beta}d^2\sigma$$

Then, $< S[x+\delta X] >$ contains a term in its second order Taylor expansion given by

$$\int B_{\mu\nu,\rho\sigma}(x)\epsilon^{\alpha\beta} < \delta X^{\rho}.\delta X^{\sigma}.\delta X^{\mu}_{,\alpha}.\delta X^{\nu}_{,\beta} > d^2\sigma$$

It also contains a term

$$\int B_{\mu\nu,\rho}(x)\epsilon^{\alpha\beta} < \delta X^{\rho}\delta X^{\mu}_{,\alpha}\delta X^{\nu}_{,\beta} > d^2\sigma$$

In order to evaluate the expectation of the product of three and four quantum string amplitude appearing above, we must first calculate approximately the string propagator in the presence of the background gauge field. That is done by using the equation of motion

$$\partial^\alpha \partial_\alpha X_\mu = K.H_{\mu\nu\rho}(X)\epsilon^{\alpha\beta} X^\nu_{,\alpha} X^\rho_{,\beta}$$

An example of the computation of the propagator in an external field. Suppose $\phi(x)$ whose Lagrangian is given by

$$L = (1/2)(\partial_t\phi)^2) - (1/2)(\nabla\phi)^2 - V\phi^2/2$$

where $V = V(x)$ is an external field. This Lagrangian describes a particle whose mass changes as it moves around in the medium. The equation of motion is

$$\partial_t^2\phi - \nabla^2\phi + V\phi = 0$$

The canonical momentum density is given by

$$\pi(t, r) = \partial L/\partial\partial_t\phi(t, r) = \partial_t\phi(t, r)$$

and hence the equal time canonical commutation relations are

$$[\phi(t, r), \partial_t\phi(t, r')] = i\delta(r - r'), [\phi(t, r), \phi(t, r')] = 0$$

Define the propagator

$$D(t, r|t', r') =< T(\phi(t, r).\phi(t', r')) >= \theta(t-t') < \phi(t, r).\phi(t', r') >$$
$$+\theta(t'-t) < \phi(t', r')\phi(t, r) >$$

We find that

$$\partial_t D = \delta(t-t') < [\phi(t, r), \phi(t, r')] > +\theta(t-t') < \partial_t\phi(t, r).\phi(t', r') >$$
$$+\theta(t'-t) < \phi(t', r')\partial_t\phi(t, r) >$$

$$= \theta(t - t') < \partial_t\phi(t, r).\phi(t', r') > +\theta(t' - t) < \phi(t', r')\partial_t\phi(t, r) >,$$
$$\partial_t^2 D = \delta(t - t') < [\partial_t\phi(t, r), \phi(t', r')] >$$
$$+\theta(t - t') < \partial_t^2\phi(t, r).\phi(t', r') > +\theta(t' - t) < \phi(t', r').\partial_t^2\phi(t, r) >$$
$$= -i\delta(t - t')\delta(r - r') + < T(\nabla_r^2 - V(t, r))\phi(t, r)).\phi(t', r')) >$$

so that

$$(\partial_t^2 - \nabla_r^2 + V(t, r))D(t, r|t', r') = -i\delta(t - t')\delta(r - r')$$

or equivalently,

$$(\Box + V)D(x|x') = -i\delta(x - x')$$

which means that formally the propagator of the field is given by

$$D(x|x') = -i(\Box + V)^{-1}$$

More generally, suppose the Lagrangian has the form

$$L = (1/2)(\partial_t\phi)^2 - (1/2)(\nabla\phi)^2 - m^2\phi^2/2 - \delta.V(t, r, \phi, \partial_t\phi, \nabla_r\phi)$$

where δ is a small perturbation parameter which shows that the external field is small. Our aim is to develop a scheme for expanding the field propagator in powers of δ and identifying the first few terms in this perturbation series. Consider the generalized KG field equation

$$\partial^2\phi_t(r) - \nabla^2\phi_t(r) + m^2\phi_t(r) + \delta.V(\phi(t,r), \nabla\phi(t,r)) = 0$$

where δ is a small perturbation parameter. Write the solution as

$$\phi(t,r) = \phi_0(t,r) + \delta.\phi_1(t,r) + O(\delta^2)$$

and then derive using first order perturbation theory

$$\partial_t^2\phi_0(t,r) - \nabla^2\phi_0(t,r) + m^2\phi(t,r) = 0$$

$$\partial_t^2\phi_1 - \nabla^2\phi_1 + m^2\phi_1 + V(\phi_0, \nabla\phi_0) = 0$$

This gives us

$$\phi_0(t,r) = (2\pi)^{-3/2}\int[(2E(P))^{-1/2}c(P)exp(-iE(P)t+iP.r)+(2E(P))^{-1/2}c(P)^* exp(iE(P)t-iP.r)]d^3P$$

where

$$E(P) = \sqrt{m^2 + P^2} = p^0$$

We then find that

$$\partial_t\phi_0(t,r) = \int[-i\sqrt{E(P)/2}c(P)exp(-ip.x) + i\sqrt{E(P)/2}c(P)^*exp(ip.x)]d^3P$$

and hence the CCR

$$[\phi_0(t,r), \partial_t\phi_0(t,r')] = i\delta^3(r-r')$$

gives us

$$[c(P), c(P')^*] = \delta^3(P-P')$$

Consider the simple linear case when $V = V(t,r) = V(x)$ does not depend on ϕ. Then we have

$$[\Box + m^2 + \delta.V]\phi = 0$$

which means that upto first order perturbation theory,

$$\phi_1(t,r) = -[\Box + m^2]^{-1}(V\phi_0) = G.(V\phi_0)$$

where

$$G(x) = -(\Box + m^2)^{-1} = \int(p^2 - m^2 + i\epsilon)^{-1}exp(ip.x)d^4p$$

is the unperturbed KG propagator. Thus,

$$\phi_1(x) = \int G(x-x')V(x')\phi_0(x')d^4x'$$

and the corrected propagator G_c upto first order is given by

$$G_c = G + \delta.GVG$$

This could have also be derived directly by using the exact expansion

$$G_c = -(\Box + m^2 + \delta.V)^{-1} = G + G.\sum_{n\geq 1} \delta^n (V.G)^n$$

The LDP problem in this context: Let V be a weak amplitude random potential. Compute the propagator for the KG particle in such a random potential and use this propagator to compute scattering amplitudes for KG particles. Calculate the rate function for these scattering probability amplitudes and hence derive approximate formulas for the probability that these scattering amplitudes will deviate from specific amplitudes by and amount greater than a given threshold.

1.7 The Atiyah-Singer Index theorem and LDP problems associated with it

The Dirac operator in curved space-time taking into account both the gravitational spin connection and the Yang-Mills connection is given by

$$D = \gamma^\mu \nabla_\mu$$

wher

$$\gamma^\mu = \gamma^a e_a^\mu(x),$$

$$\nabla_\mu = \partial_\mu + ieA_\mu + \Gamma_\mu$$

where e_a^μ is a tetrad basis of our curved space-time, A_μ and Γ_μ being respectively the Yang Mills connection and the gravitational spin connection. The index of this Dirac operator is given by

$$Ind(D) = str(exp(-\beta D^2))$$

where β is any real number which we will later on force it to tend to zero and str denotes supertrace. To clarify the exact structure of this Dirac operator, we write it as

$$D = \begin{pmatrix} 0 & D_0 \\ D_0^* & 0 \end{pmatrix}$$

where this partitioning corresponds to the partitioning of the 4×4 Dirac Gamma matrices into 2×2 blocks. Thus,

$$D^2 = \begin{pmatrix} D_0 D_0^* & 0 \\ 0 & D_0^* D_0 \end{pmatrix}$$

so that

$$str(exp(-\beta D^2)) = Tr(exp(-\beta D_0 D_0^*)) - Tr(exp(-\beta D_0^* D_0))$$

To calculate this supertrace, we express the Dirac Gamma matrices in terms o Clifford algebra creation and annihilation elements:

$$\gamma^a = e^a + e_a$$

where

$$e^a \omega = e^a \wedge \omega,$$

and e_a is the adjoint of e^a. We have the obvious Clifford algebra anticommuta tion relations

$$e^a e_b + e_b e^a = \delta^a_b, e^a e^b + e^b e^a = 0, e_a e_b + e_b e_a = 0$$

and hence,

$$\gamma^a \gamma^b + \gamma^b \gamma^a = \eta^{ab}$$

We then introduce the Getzler scaling operator on the space of exterior forms:

$$S(\epsilon)\omega = \epsilon^{deg(\omega)}\omega$$

Then, we get

$$S(\epsilon)(\gamma^a.\omega) = S(\epsilon)(e^a \wedge \omega + e_a(\omega))$$

$$= \epsilon^{deg(\omega)+1} e^a \wedge \omega + \epsilon^{deg(\omega)-1} e_a(\omega)$$

$$= \epsilon e^a \wedge S(\epsilon)(\omega) + \epsilon^{-1} e_a S(\epsilon)(\omega)$$

which is the same as saying that

$$S(\epsilon)\gamma^a.S(\epsilon)^{-1} = \epsilon e^a + \epsilon^{-1} e_a$$

This means that

$$lim_{\epsilon\to\infty} \epsilon^{-1} S(\epsilon)\gamma^a S(\epsilon)^{-1} = e^a$$

This idea can be used to reduce computations regarding supertrace into exterior integration on a manifold and hence to relate the index of the Dirac operator defined as a supertrace to an integral of forms on the Riemannian manifold.

The LDP problem associated with the Atiyah-Singer index theorem: Given the Dirac operator in curved space-time with a Yang-Mills non-Abelian gauge connection potential $A_\mu(x)$, we note that it can be expressed as

$$D = \gamma^a e^\mu_a(x)(\nabla_\mu + ieA_\mu)$$

where

$$\nabla_\mu = \partial_\mu + \Gamma_\mu$$

is the curved space-time covariant derivative in the spinor representation. Compute the index of D^2 in the situation when the gauge potential has random terms. How much should the gauge potential vary by so that the Dirac index changes from one integer to another ?

1.8 LDP problems in general relativity

[1] Parametrize the metric of space-time by a finite parameter set. Set up the differential equations for geodesic motion for a charged particle in this metric in the presence of a noisy electromagnetic field:

$$g_{\mu\nu} = g_{\mu\nu}(x|\theta)$$

$$\frac{d^2x^\mu(\tau)}{d\tau^2} + \Gamma^\mu_{\alpha\beta}(x(\tau)|\theta)\frac{dx^\alpha(\tau)}{d\tau}.\frac{dx^\beta(\tau)}{d\tau} = \sqrt{\epsilon}F^{\mu\nu}(x|\theta)\frac{dx^\nu(\tau)}{d\tau}$$

Solve these geodesic equations using perturbation theory and assuming the electromagnetic field to be a zero mean Gaussian random field:

$$A_\mu(x|\theta) = A^\nu(x)g_{\mu\nu}(x|\theta),$$

$$F_{\mu\nu}(x|\theta) = A_{\nu,\mu}(x|\theta) - A_{\mu,\nu}(x|\theta),$$

$$F^{\mu\nu}(x|\theta) = g^{\mu\alpha}(x|\theta)g^{\nu\beta}(x|\theta)F_{\alpha\beta}(x|\theta)$$

and then calculate the rate function for the particle worldline $x^\mu(\tau)$. Hence determine the approximate probability that the particle worldline will deviate from a given non-random trajectory by an amount more than a given threshold ϵ and control the parameter vector θ so that this probability of deviation is a minimum. Repeat for different kinds of statistics for the random electromagnetic field.

[2] Consider the string field X interacting with a scalar field $\Phi(X)$, the metric tensor g of space-time having curvature $R(X)$, and an external antisymmetric second rank gauge potential $B(X)$ with the total action being given by

$$S[X] = S_1[X] + S_2[X] + S_3[X]$$

where

$$S_1[X] = (1/2)\int g_{\mu\nu}(X)h^{\alpha\beta}\sqrt{-h}\partial_\alpha X^\mu.\partial_\beta X^\nu d^2\sigma$$

$$S_2[X] = \int R(X)\Phi(X)\sqrt{-h}d^2\sigma$$

$$S_3[X] = \int B_{\mu\nu}(X)\epsilon^{\alpha\beta}\partial_\alpha X^\mu\partial_\beta X^\nu\sqrt{-h}d^2\sigma$$

where $h_{\alpha\beta}(\tau,\sigma)$ is the string world sheet metric. We ask the question, what should the field equations for Φ, g, B be so that the total action $S[X]$ has conformal invariance after taking quantum averages around the classical centre of mass of the string. We have already noted that in the presence of the gauge field alone, the string field satisfies

$$\Box X_\mu = H_{\mu\nu\rho}(X)dX^\nu \wedge dX^\rho$$

which means that the approximate corrected local string propagator around the classical centre of mass x^μ of the string is given by

$$D_{\mu\nu}(\tau,\sigma|\tau',\sigma') =$$

$$D^{(0)}_{\mu\nu}(\tau,\sigma|\tau',\sigma')+H_{\mu\alpha\beta}(x)H_{\nu\rho\sigma}(x)\epsilon^{ab}\epsilon^{cd}$$

$$< \delta X^\alpha_{,a}(\tau,\sigma)\delta X^\beta_{,b}(\tau,\sigma)\delta X^\rho_{,c}(\tau',\sigma')\delta X^\sigma_{,d}(\tau',\sigma') >$$

where

$$\delta X^\mu(\tau,\sigma) = X^\mu(\tau,\sigma) - x^\mu$$

We next observe that the average (quantum average) change in $S_3[X]$ due to quantum fluctuations of the string around its classical C.M position x^μ is approximately given by

$$< \delta S_3[X] >= \int B_{\mu\nu,\rho}(x) < \delta_1 X^\rho(\tau,\sigma).\epsilon^{ab}X^\mu_{,a}(\tau,\sigma)X^{nu}_{,b}(\tau,\sigma) > d^2\sigma$$

where now $\delta_1 X^\mu$ is the change in X^μ caused by the presence of the gauge field and is approximately given by

$$\delta_1 X_\mu = D^{(0)}.(H_{\mu\nu\rho}(X)dX^\nu \wedge dX^\rho)$$

$$\approx H_{\mu\nu\rho}(x)\epsilon^{ab}D^{(0)}.(X^\nu_{,a}.X^\rho_{,b})$$

It follows therefore, that the average change in S_3 locally at the classical point x due to small quantum string fluctuations is a quadratic form in $H_{\mu\nu\rho}(x)$ and is given by

$$< \delta S_3[X] >= B_{\mu\nu,\rho}(x)H^\rho_{\alpha\beta}(x)\epsilon^{ab}\epsilon^{cd}\int [D^{(0)}(\tau,\sigma|\tau'\sigma') < \delta X^\alpha_{,a}(\tau',\sigma')\delta X^\beta_{,b}(\tau',\sigma')$$

$$.\delta X^\mu_{,c}(\tau,\sigma)\delta X^\nu_{,d}(\tau,\sigma) > d^2\sigma.d^2\sigma']$$

Using the standard representation of the unperturbed string propagator, this expression is easily seen to be of the form $K.H_{\mu\nu\rho}(x)H^{\mu\nu\rho}(x)$. Likewise, we find that the average change in $S_1[X]$ due to local quantum fluctuations of the string field has the form

$$< \delta S_1[X] >= \int (1/4)h^{\alpha\beta}\sqrt{-h}(g_{\mu\nu,\rho\sigma}(x) < \delta X^\rho.\delta X^\sigma X^\mu_{,\alpha}\delta X^\nu_{,\beta} > d^2\sigma$$

and in a normal coordinate system this can be expressed as $K.R(x)$ where K is a constant and $R(x)$ is the Riemann curvature scalar of the metric g at the classical point x. We can also calculate higher order corrections to the change in the string action caused by gravitational corrections to the string propagator. Such corrections will contain quadratic and higher order terms in the Riemann curvature tensor like $R_{\mu\nu\alpha\beta}(x)R^{\mu\nu\alpha\beta}(x)$ and $R(x)^2$. Let us give a sample calculation for this: The equation of string motion in the gravitational field is given by

$$\eta^{\alpha\beta}(g_{\mu\nu}(X)X^\nu_{,\beta})_{,\alpha} = \eta^{\alpha\beta}g_{\rho\nu,\mu}(X)X^\rho_{,\alpha}X^\nu_{,\beta}$$

which is the same as

$$g_{\mu\nu}(X)\Box X^\nu + \eta^{\alpha\beta}[g_{\mu\nu,\rho}X^\rho_{,\alpha}X^\nu_{,\beta} - g_{\rho\nu,\mu}(X)X^\rho_{,\alpha}X^\nu_{,\beta}]$$

$$\Box X_\mu = \Gamma_{\mu\nu\alpha}(X)\epsilon^{ab}X^\nu_{,a}X^\alpha_{,b}$$

Chapter 2

LDP in Biology, Neural Networks, Electromagnetic Measurements, Cosmic Expansion

2.1 The Importance of Mathematical Models in Medicine

Since the time of Newton, differential equations have been used to model a wide range of physical phenomena starting from planetary motion and going on upto controlling the flight of rockets and satellites in space, modeling fluid motion, electromagnetic fields in space-time, circuit dynamics, describing the dynamics of elementary particles that participate in absorption and scattering processes in the subatomic world. Differential equations are used in technology to model the dynamics of speech and image data in time, space and in space-time. These models enable us to construct compression algorithms and thus build speech and image synthesizers. More recently, complex phenomena involving innumberable degrees of freedom like galactic evolution, turbulent motion of fluids, thermal noise generated in electronic circuits and even the dynamics of the vocal tract that generates speech and evolving image and electromagnetic fields subject to noise have been successfully modeled by combining the theory of ordinary and partial differential equations with the theory of probability and stochastic processes resulting in the general theory of stochastic differential equations. Any model of this sort usually contains a finite number of unknown parameters which are estimated/trained using sample input-output data. After such a training process, the trained model can be used to predict new phenomena like the output of a system when the input is different from the set used to train the model. Apart from this application, the trained model for each class of input output data yields parameter sets which can be used for classifying the natur of the system.

Recently, mathematical models have been used to model the dynamics c phenomena taking place within the human body like the onset of disease in th blood, bones, nervous system and in the brain. All these models of the huma body can be cast into the form of a stochastic differential system with unknow parameters excited by an input process and generating an output process. Fo example, consider the case of a diseased nervous system involving the brain When we give a certain stimulus to the senses, the brain generates a certai kind of EEG signal field and the statistics of this signal field will depend upo the input stimulus, upon the brain parameters and also upon the statistics c the noise that one incorporates into the brain dynamics while modeling it. B modeling the EEG signal field using a partial differential equation in time an space with stochastic noise included and with the pde coefficients being fre control parameters, we can, from given input sensory and output EEG data estimate the control parameters either using block processing techniques lik least squares, maximum likelihood methods or using real time estimators lik the extended Kalman filter (EKF). Block processing techniques are useful whe the full future data is available at any given time and when the parameters d not vary with time. On the other hand, real time techniques are useful when th future data is not available at any given time and when the parameters fluctuat slowly with time (Tracking of non-stationarities). In all of these parameters, on matches the measured output data in some sense to that computed theoreticall from the parametric model using an appropriate distance measure that usuall depends upon the statistics of the noise in the parametric differential equatio model. Each kind of brain disease will determine a different class of paramete set and these distinct and widely separated classes of parameter sets can be use to determine the class of brain disease in a fresh patient from input-output dat based on sensory input stimulus and measured output EEG data. This schem can also be used to synthesize the EEG data from some other lower dimensiona data by exploiting the correlations between the two kinds of data.

Consider another example, namely the case of blood disease. The motion o the blood in the presence of an external electromagnetic field can be modeled b the standard partial differential equations of MHD (Magneto-hydro-dynamics of conducting fluids taking into account stochastic noise. For a given applie electromagnetic field, we can measure the blood fluid velocity and density fiel at a finite number of spatial points and then use a nonlinear filtering algorithm to match these measured field quantities to that theoretically predicted from th stochastic pde dynamical model of the blood and thereby estimate the parame ters in this model like the viscosity field and also the blood velocity and densit field values at all the spatial points where no measurement has been carrie out, ie, we solve the problem of extrapolation. The estimated blood parameter enable us as in the EEG case, to classify the nature of blood disease.

In many situations, two kinds of data on the human body, one higher dimen sional and two lower dimensional, are correlated and hence can be modeled us ing differential equations with correlated parameters. We train these differential equations using both the lower and higher dimensional data and thus estimate these parameters and arrange these in different widely separated classes. When a fresh patient is then presented with only lower dimensional data available, we can estimate his parameters using this available data either using a block processing or a real time filtering algorithm to match the available measured data to that generated by the model. Then by exploiting the correlations between the parameters of the lower and higher dimensional data, we can hope to synthesize the higher dimensional data from its differential equation model. Alternately, we can classify the nature of the disease by matching the parameters of a fresh patient estimated using only his lower dimensional data to the parameter classes obtained using the training scheme based upon both lower and higher dimensional data. In short, mathematical models of biological phenomena based on differential equations, probability theory and matrix algebra can be used to model, predict and classify the onset of newer kinds of disease by exploiting correlations. A typical example of this is the correlation between the slurring of speech and the EEG of a patient having some kind of brain disease.

The method of modeling physical and biological systems as black boxes with parameters estimated from input-output data (with for example, the black box being a stochastic differential equation or a stochastic pde with unknown coefficients) has been in recent times replaced by a neural network whose weights are trained from a given training set of i/o data and then when a fresh input is presented, the network computes the corresponding output. The neural network can be made recurrent so that its current state depends on the previous state and weights. This enables one to model the evolution of phenomena having memory.

Recently the concept of a quantum neural network (QNN) has been proposed for dynamically estimating the probability distribution of the states whose time samples describe the evolving physical phenomena. The chief equation here is the Schrodinger wave equation which governs the dynamics of the wave function whose modulus square is always a normalized probability density and hence by controlling the potential with neural weights so that its pdf (probability density function) tracks a given pdf, we are able to train the evolving weights so that it generates a given sequence of pdf's. When the QNN is presented with a fresh sequence of pdf's, the resulting weights determine the class of disease in which this pdf sequence falls. The idea of a QNN is very useful when the phenomena are random so that they are characterized not by a state sequence but by a pdf sequence which can be determined by local empirical distributions based on local time averages.

Finally, I would like to conclude by a problem involving the application of electromagnetic field theory to the problem of characterizing bone disease. The nature of bone disease is characterized by its permittivity-permeability-conductivity field which can be estimated by estimating the statistics of the scattered radiation fields when an incident random radiation field falls on the bone. The statistics of the scattered field is matched to that theoretically calculated in terms of the inhomogeneous permittivity-permeability-conductivity fields using the Maxwell equations in such a medium. Once the permittivity and permeability fields have been characterized, we train a neural network that outputs the disease classification index for a given input permittivity-permeability field. Again this scheme is simply a pde based model for the bone with the the pde being the system of Maxwell equations, the input being the incident radiation field, the output being the scattered radiation field and the system parameters being the bone permittivity-permeability field.

I hope that I have in this talk successfully conveyed the importance of using differential equations in modeling biological systems and applying these models to characterizing and predicting new diseases.

2.2 LDP related problems in neural networks and artificial intelligence

Let $\mathbf{X}(t)$ be the input process (assumed to be non-random) and let $\mathbf{Y}(t)$ be the output process. Assume that the true plant model is define by differential equation

$$d\mathbf{Y}(t) = \mathbf{F}(t, \mathbf{Y}(t), \mathbf{X}(t)|\theta)dt + \mathbf{G}(t, \mathbf{Y}(t), \mathbf{X}(t)|\theta)d\mathbf{B}(t)$$

where $\mathbf{B}(.)$ is a vector valued Brownian motion process. Our NN based method for estimating θ is as follows. We given an input $\mathbf{X}$ to the plant, measure the output $\mathbf{Y}$ and from this i/o data, construct the MLE $\hat{\theta}$ of the plant using standard methods in sde theory:

$$\hat{\theta}(T) = argmin_\theta \int_0^T (d\mathbf{Y}(t) - \mathbf{F}(t, \mathbf{Y}(t), \mathbf{X}(t)|\theta)dt)^T$$

$$.(dt.\mathbf{G}(t, \mathbf{Y}(t), \mathbf{X}(t)|\theta).(\mathbf{G})(t, \mathbf{Y}(t), \mathbf{X}(t)|\theta)^T)^{-1}(d\mathbf{Y}(t) - \mathbf{F}(t, \mathbf{Y}(t), \mathbf{X}(t)|\theta)dt)$$

We then train our NN by changing the plant parameter θ, estimating it as $\hat{\theta}$ and feeding the o/p plant process $\mathbf{Y}$ to the NN as its input and matching the NN output to the MLE $\hat{\theta}$ of θ. In this scheme, the input process is assumed to be fixed. The trained weights will then output a sub-optimal estimate of θ for any choice of the true plant parameter based on the output generated by it. The LDP problem can then be formulated as follows: Choose and fix a value of the plant parameter θ and generate its estimate $\hat{\theta}$ based on the above training scheme with the output data used for training consisting of N independent output realizations, each of duration T. For the training scheme, let $\theta_1, ..., \theta_N$ denote the chosen values of the parameter θ and let $\mathbf{Y}_1, ..., \mathbf{Y}_N$ denote the corresponding independent o/p processes used to train the NN weight. The trained weight vector is denoted by $\mathbf{W}(\mathbf{Y}_1, ..., \mathbf{Y}_N, \theta_1, ..., \theta_N)$. Note that each $\mathbf{Y}_j$ is of duration T and is generated by an independent Brownian motion process. Then for a given plant parameter θ that does not belong to the training set values, we generate an independent output process $\mathbf{Y}$ and use the NN to estimate θ. Denote this estimate by $\hat{\theta}$. Thus with $\mathbf{H}$ denote the NN nonlinear function, we can write

$$\hat{\theta} = \mathbf{H}(\mathbf{W}(\mathbf{Y}_1, ..., \mathbf{Y}_N, \theta_1, ..., \theta_N), \mathbf{Y})$$

The problem is to calculate the probability $P(\| \hat{\theta} - \theta \| > \delta)$ using the theory of large deviations as a function of $N, T, \theta_1, ..., \theta_N, \theta$ and the given input process. It should be noted that the Brownian motion processes $\mathbf{B}_1, ..., \mathbf{B}_N, \mathbf{B}$ used in the generation of $\mathbf{Y}_1, ..., \mathbf{Y}_N, \mathbf{Y}$ are independent realizations.

2.3 LDP related problems to cosmic expansion in general relativity

Consider the RW metric of homogeneous and isotropic space-time. Suppose we consider first order perturbations to the Einstein field equations around this background metric. In these equations, we take into account the extra energy-momentum tensor correction terms coming from viscous and heat terms. We incorporate small random terms into this energy-momentum tensor like for example random perturbations of the viscosity coefficients and the the heat diffusion coefficients. The resulting metric perturbations and velocity and density perturbations satisfy a set of linear pde's. We now take into account a low amplitude cosmic microwave background radiation field described as a homogeneous and isotropic zero mean Gaussian field with spatio-temporal correlations and set up the linearized Einstein field equations in such a background. The resulting field equations will involve quadratic functionals of the electromagnetic background field because the energy-momentum tensor of the electromagnetic field is a quadratic function of the electromagnetic field tensor. Thus, the driving force for the metric, velocity and density perturbations will be the initial conditions as well as a generalized χ-square field and it is an important problem to derive the rate functional for these perturbations in terms of the rate functional of the generalized χ-square field.

2.4 LDP problems in biology

Aim: To detect the onset of osteoporosis (brittleness of the bones) by measuring the statistics of the electromagnetic radiation field scattered by the bone medium. Let $\epsilon(\omega, \mathbf{r})$ and $\mu(\omega, \mathbf{r})$ denote respectively the permittivity and permeability of the bone medium and let $\mathbf{E}_i(\omega, \mathbf{r})$ and $\mathbf{H}_i(\omega, \mathbf{r})$ denote the incident electromagnetic field on the bone. Let $\mathbf{E}_s(\omega, \mathbf{r}), \mathbf{H}_s(\omega, \mathbf{r})$ denote the scattered fields so that the total electromagnetic field

$$\mathbf{E} = \mathbf{E}_i + \delta.\mathbf{E}_s + O(\delta^2), \mathbf{H} = \mathbf{H}_i + \delta.\mathbf{H}_s + O(\delta^2)$$

satisfies the Maxwell equations

$$curl\mathbf{E}(\omega, \mathbf{r}) = -j\omega\mu(\omega, \mathbf{r})\mathbf{H}(\omega, \mathbf{r})$$

$$curl\mathbf{H}(\omega, \mathbf{r}) = j\omega\epsilon(\omega, \mathbf{r})\mathbf{E}(\omega, \mathbf{r})$$

We write

$$\epsilon(\omega, \mathbf{r}) = \epsilon_0\mathbf{I}_3 + \delta.\epsilon\chi_e(\omega, \mathbf{r}),$$

$$\mu(\omega, \mathbf{r}) = \mu_0\mathbf{I}_3 + \delta.\mu_0\chi_m(\omega, \mathbf{r})$$

where δ is a small perturbation parameter and $\chi_e(\omega, \mathbf{r}), \chi_m(\omega, \mathbf{r})$ are 3×3 complex matrix valued functions of frequency and spatial location which correspond respectively to a small inhomogeneous and anisotropic perturbation to the permittivity and permeability. Our understanding is that estimates of these two matrix valued functions would be enough for us to characterize the nature of the bone disease. We choose a set of 3×3 matrix valued test/basis functions $\mathbf{A}_k(\omega, \mathbf{r}), k = 1, 2, ..., p$ and assume that these inhomogeneous and anisotropic electric and magnetic susceptibilities can be expanded as a linear combination of these matrix test functions:

$$\chi_e(\omega, \mathbf{r}) = \sum_{k=1}^{p} \theta[k]\mathbf{A}_k(\omega, \mathbf{r}),$$

$$\chi_m(\omega, \mathbf{r}) = \sum_{k=1}^{p} \phi[k]\mathbf{A}_k(\omega, \mathbf{r}),$$

substitute these into the Maxwell equations and write down the equations upto $O(\delta)$:

$$curlcurl\mathbf{E} = -j\omega curl(\mu\mathbf{H}), curlcurl\mathbf{H} = j\omega curl(\epsilon\mathbf{E}),$$

$$div(\epsilon\mathbf{E}) = 0, div(\mu\mathbf{H}) = 0$$

$$curlcurl\mathbf{E}_s = -j\omega.curl(\mu_0\mathbf{H}_s + \mu_0\chi_m.\mathbf{H}_i)$$

$$curlcurl\mathbf{H}_s = j\omega curl(\epsilon_0\mathbf{E}_s + \epsilon_0\chi_e.\mathbf{E}_i)$$

where

$$curl\mathbf{E}_s = -j\omega\mu_0\mathbf{H}_s - j\omega\mu_0\chi_m\mathbf{H}_i,$$

$$curl\mathbf{H}_s = j\omega\epsilon_0\mathbf{E}_s + j\omega\epsilon_0\chi_e\mathbf{E}_i,$$

$$div\mathbf{E}_s + div(\chi_e.\mathbf{E}_i) = 0,$$

$$div\mathbf{H}_s + div(\chi_m\mathbf{H}_i) = 0$$

where

$$k^2 = \omega^2\epsilon_0\mu_0$$

These first order perturbation equations can be solved for $\mathbf{E}_s, \mathbf{H}_s$ using standard matrix Green's functions in the form:

$$\mathbf{E}_s(\omega, \mathbf{r}) = \int [\mathbf{G}_{11}(\omega, \mathbf{r}-\mathbf{r}')(\chi_e(\omega, \mathbf{r}').\mathbf{E}_i(\omega, \mathbf{r}')) + \mathbf{G}_{12}(\omega, \mathbf{r}-\mathbf{r}')(\chi_m(\omega, \mathbf{r}').\mathbf{H}_i(\omega, \mathbf{r}')]d^3r',$$

$$H_s(\omega,\mathbf{r}) = \int [\mathbf{G}_{21}(\omega,\mathbf{r}-\mathbf{r}')(\chi_e(\omega,\mathbf{r}').\mathbf{E}_i(\omega,\mathbf{r}'))+\mathbf{G}_{22}(\omega,\mathbf{r}-\mathbf{r}')(\chi_m(\omega,\mathbf{r}').\mathbf{H}_i(\omega,\mathbf{r}')]d$$

Equivalently, in terms of the coefficient parameters $\theta[k], \phi[k]$,

$$\mathbf{E}_s(\omega,\mathbf{r}) = \sum_k \theta[k] \int \mathbf{G}_{11}(\omega,\mathbf{r}-\mathbf{r}')\mathbf{A}_k(\omega,\mathbf{r}').\mathbf{E}_i(\omega,\mathbf{r}')d^3r'$$

$$\sum_k \phi[k] \int \mathbf{G}_{12}(\omega,\mathbf{r}-\mathbf{r}')\mathbf{A}_k(\omega,\mathbf{r}').\mathbf{H}_i(\omega,\mathbf{r}')d^3r'$$

$$\mathbf{H}_s(\omega,\mathbf{r}) = \sum_k \theta[k] \int \mathbf{G}_{21}(\omega,\mathbf{r}-\mathbf{r}')\mathbf{A}_k(\omega,\mathbf{r}').\mathbf{E}_i(\omega,\mathbf{r}')d^3r'$$

$$\sum_k \phi[k] \int \mathbf{G}_{22}(\omega,\mathbf{r}-\mathbf{r}')\mathbf{A}_k(\omega,\mathbf{r}').\mathbf{H}_i(\omega,\mathbf{r}')d^3r'$$

By taking measurements of the scattered radiation fields at a finite set of pixel points, we can estimate using the least squares method the parameters $\theta[k], \phi[k], k = 1, 2, ..., p$. Each degree of osteoporosis will correspond to a definite class of the parameter set and by matching a given parameter set to such classes, we can determine the degree of the onset of the disease.

The LDP problem: When the inhomogeneous and anisotropic perturbations to the permittivity and permeability fields are weak amplitude Gaussian/non-Gaussian fields with known statistics, it would follow that the coefficients appearing in their expansion w.r.t a basis set of functions will also be weak amplitude random with known statistics and hence their estimates using the above mentioned least squares method will also be random and in principle, expressible as functions of the true random parameters. When the number of measurement points is large, the question is what is the rate function of these parameters ? More generally, when the permittivity and permeability fields are small random perturbations of known non-random fields, then the estimates of the coefficients in their basis expansions using the least squares or maximum likelihood method will also be small random perturbations of the non-random coefficients appearing in the basis function expansions of the non-random components of the permittivity and permeability fields. The question is what is the rate function for small random perturbations in terms of the statistics of the random perturbations of the permittivity and permeability fields and whether we can control the parameters of the non-random components so that the probability of deviation of these coefficient estimates from the true non-random values by an amount more than a threshold will be minimized.

2.5 A Sensitive Quantum Mechanical Method for Measuring the scattered Electromagnetic Fields

2.5 A Sensitive Quantum Mechanical Method for Measuring the scattered Electromagnetic Fields

Assume that the electromagnetic field consists of a strong non-random component and a weak zero mean random component. We can thus express th electromagnetic potentials as

$$A_\mu = A_\mu^{(0)}(x) + \delta A_\mu(x,\epsilon)$$

where ϵ is a small parameter so that δA_μ converges to the zero random field a $\epsilon \to 0$. Assume that δA_μ has an LDP rate function $I(A)$ over a region of space time. Now when this random electromagnetic field falls on a Dirac electro bound to its nucleus, the perturbation to the Hamiltonian becomes

$$\Delta H(x) = e(\alpha, \mathbf{A}(x)) - e\Phi(x) + e(\alpha, \delta\mathbf{A}(x,\epsilon)) - e\delta\Phi(x,\epsilon) = \Delta H_0(x) + \delta H_0(x,\epsilon$$

where

$$\Delta H_0(x) = e(\alpha, A(x)) - e\Phi(x)$$

is the perturbation to the Dirac Hamiltonian caused by the non-random component of the electromagnetic four potential and $\delta H_0(x,\epsilon)$ is the perturbatio caused by the random component of the electromagnetic four potential. Not that

$$A(x) = (A^1, A^2, A^3), \Phi(x) = A^0$$

Compute using first order perturbation theory, the approximate transition prob ability of the Dirac electron in time $[0,T]$ caused by the non-random componen ΔH_0 and the small random change in this transition probability caused by th random component $\delta H_0(x,\epsilon)$ and hence derive an expression for the LDP rat function for this random shift in the transition probability and then try t introduce control parameters into the non-random components of the electro magnetic fields and control these parameters so that the probability that th shift in the transition probability caused by the random component $\delta H(x,\epsilon$ exceeding a given threshold is minimized.

Chapter 3

LDP in Signal Processing, Communication and Antenna Design

3.1 Review

The paper is about dual band real time signal processing implemented using microwave engineering. It further discusses experimental verification of dual band edge detectors. I have the following questions:

[a] SSB (Single sideband, ie, USB (Upper sideband) and LSB(lower sideband)) modulation at microwave frequencies need to be implemented in analog communication applications. For doing this, we require the continuous time Hilbert transformer. Let $x(t)$ be a signal and $\hat{x}(t)$ its Hilbert transform. We write

$$y(t) = x(t) + j\hat{x}(t), z(t) = x(t) - j\hat{x}(t)$$

These have Fourier transforms

$$Y(\omega) = (1 + sgn(\omega))X(\omega), Z(\omega) = (1 - sgn(\omega))X(\omega)$$

Thus, if $x(t)$ is a real lowpass signal with bandwidth much smaller than $2\omega_0$ where ω_0 is the carrier frequency, then the USB signal is given by

$$\phi_u(t) = y(t).exp(j\omega_0 t) + z(t)exp(-j\omega_0 t)$$

and the LSB signal is given by

$$\phi_l(t) = y(t)exp(-j\omega_0 t) + z(t)exp(j\omega_0 t)$$

Equivalently, in the frequency domain

$$\Phi_u(\omega) = Y(\omega - \omega_0) + Z(\omega + \omega_0),$$

$$\Phi_l(\omega) = Y(\omega + \omega_0) + Z(\omega - \omega_0)$$

If these operations are performed using the dual band microwave methods using transmission line elements suggested in the paper, then what errors will occur in these formulas ? Further, when this implementation is performed at two different centre frequencies and the corresponding spectra overlap, then how to recover the low pass signal from the USB and the LSB components ?

[b] Some clarity is required in specifying how the $sgn(\omega)$ transfer function of the Hilbert transfvormer can be generated in the continuous frequency domain. For discrete set of frequencies, the microwave implementation is able to implement this by simply phase shifting the positive frequencies by $-\pi/2$ and the negative frequencies by $-\pi/2$.

[27] gives the Hilbert transform of a sinuosidal carrier signal by a periodic pulse by simply phase shifting the discrete positive spectral components by $-\pi/2$ and the negative spectral components by $\pi/2$. If however, the periodic pulse is replaced by an aperiodic lowpass signal, then the Hilbert transform of $x(t)cos(\omega_0 t)$ would be given by $x(t).sin(\omega_0 t)$ or equivalently in the frequency domain,

$$x(t).cos(\omega_0 t) \rightarrow_F X(\omega - \omega_0)/2 + X(\omega + \omega_0)/2$$

$$x(t).sin(\omega_0 t) \rightarrow -jX(\omega - \omega_0)/2 + jX(\omega + \omega_0)/2$$

If some part of the spectrum of $x(t)$ becomes greater than ω_0, then this formula would break down. Is it possible to suggest an algorithm to make the microwave Hilbert transformer robust against such a leakage ?

[c] If $x(t)$ is a stationary random bandpass signal then its complex lowpass envelope is given by $\tilde{x}(t) = (x(t) + j\hat{x}(t)).exp(j\omega_0 t)$. The power spectral density of $\tilde{x}(t)$ is easily seen to be given by

$$S_{\tilde{x}}(\omega) = 2(1 + sgn(\omega - \omega_0))S_x(\omega - \omega_0)$$

If the Hilbert transformer used in the paper based on microwave elements is applied to such a random signal, then what will be the error in the spectrum of the complex lowpass envelope ?

[d] For two dimensional signals, is it possible to design using microwave elements the Hilbert transformer ?

After addressing the above commments briefly, I recommend publication of the paper.

3.2 Large deviation problems in SSB modulation

Let $x_\epsilon(t), t \in \mathbb{R}, \epsilon \rightarrow 0$ be a family of stationary random processes. Assume that the rate function of this family is $I(x)$:

$$Pr((x_\epsilon(t) : t \in [0, T]) \in B) \approx exp(-inf(I(x) : x \in B)/\epsilon), \epsilon \rightarrow 0$$

where B is a Borel subset of $\mathbb{R}^{[0,T]}$. Then compute the rate function of the Hilbert transform of x_ϵ and also of its USB and LSB versions.

3.3 What is meant by estimating a quantum field in space-time

A quantum field in space-time is a family of operator valued functions $X(t, \mathbf{r})$ of time and space arguments that evolves according to a noisy Hamiltonian $H(t)$. Its Heisenberg dynamics can be expressed as

$$-ih\partial X(t, \mathbf{r}))/\partial t = [H(t), X(t, \mathbf{r})]$$

We can formally write the noisy Hamiltonian as

$$H(t) = H_0(t) + \sum_k (L_{k,1}(t) + L_{k,2}(t)(-1)^{\Lambda(t)}) \otimes W_k(t))$$

where $H_0(t), L_{k,a}(t), a = 1, 2$ are system space operators and $W_k(t)$ are quantum white noise processes assuming values in the tensor product of a Boson and a Fermion Fock space. The noisy dynamics is characterized by an evolution operator $U(t)$ that satisfies Schrodinger's equation

$$ihU'(t) = H(t)U(t)$$

or more precisely, in the quantum Ito differential notation of Hudson and Parthasarathy, writing $W_k(t) = B'_k(t)$ and with $H_0(t)$ containing the quantum Ito correction terms,

$$ihdU(t) = (H_0 dt + \sum_k (L_{k,1}(t) + L_{k,2}(t)(-1)^{\Lambda(t)}) \otimes dB_k(t))U(t)$$

where the $B_k(t)'s$ are the fundamental noise processes of the Hudson-Parthasarathy quantum stochastic calculus. $\Lambda(t)$ is a quantum Poisson process and the factor of $(-1)^{\Lambda(t)}$ accounts for the fact that the we can have even Fermionic noise components. The problem of estimating the entire quantum field $X = X(t, \mathbf{r}), \mathbf{r} \in \mathbb{R}^3$ at a given time t does not carry meaning since these observables at any given time and at different spatial points do not commute. So to make the problem more meaningful, we must project this entire quantum field onto an Abelian Von-Neumann algebra of operators and then jointly estimate this projected field. More generally, we can consider super noise processes as follows. Let $\Lambda^a_b(t)$ be the fundamental noise processes satisfying the quantum Ito formula

$$d\Lambda^a_b(t).d\Lambda^c_d(t) = \epsilon^a_d d\Lambda^c_b(t), a, d \geq 0$$

where $\Lambda^0_0(t) = t$ and ϵ^a_b is one iff $a = b \geq 1$ otherwise it is zero. Then, we define

$$\Lambda(t) = \sum_{k=1}^{r} \Lambda^a_a(t)$$

and define

$$G(t) = (-1)^{\Lambda(t)}$$

The super-noise processes are defined by

$$\xi_b^a(t) = \int_0^t G(s)^{\sigma_b^a} d\Lambda_b^a(s) ds$$

where σ_b^a is 0 iff either $1 \leq a, b \leq r$ or $r+1 \leq a, b \leq N$ otherwise it is 1. Thus, $(((-1)^{\sigma_b^a}))$ is the parity matrix. It is one for even matrices, ie matrices having the block structure $diag[A, B]$ where $A \in \mathbb{C}^{r \times r}$ and $B \in \mathbb{C}^{N-r \times N-r}$ and -1 for matrices of the form

$$\begin{pmatrix} 0 & C \\ D & 0 \end{pmatrix}$$

where $C \in \mathbb{C}^{r \times N-r}, D \in \mathbb{C}^{N-r \times r}$. T.Eyre [1] has established the supercommutation rules for such super-noise processes. In order to measure the quantum field at different points in space sequentially in time, we use the Belavkin filter. First, we estimate the field at a given point r_1 in space over a time duration $[0, t_1]$ using a non-demolition family of output measurements. In other words, we estimate $j_t(X(0, r_1)) = X(t, r_1)$ as $\pi_t(X(0, r_1))$. Then, we again use the same non-demolition measurement to estimate the field at another spatial point r_2 over the time interval $[t_1, t_2]$ and so on. The collapse postulate following each measurement does not apply because the measurements are non-demolition.

3.4 Write down the noisy Schrodinger equation for an N particle system in the formalism of Hudson and Parthasarathy and derive by partial tracing, the approximate nonlinear stochastic Boltzmann equation for the state evolution of a single particle (evolution of the marginal density with noise)

This can be obtained as follows:

$$dU(t) = -i(H_0(t) + H_1(t))U(t)$$

where $H_0(t)$ is the noiseless system Hamiltonian with Ito correction terms taken into account while $H_1(t)$ is the noisy Hamiltonian of the form $\sum_k (L_k(t) \otimes W_k(t))$. The initial state of the system is $\rho(0) = \rho_s(0) \otimes \rho_b(0)$ where $\rho_s(0)$ is the initial system state and $\rho_b(0)$ is the initial bath state. After time t, the state of the system $\rho_s(t)$ is given by

$$\rho_s(t) = Tr_2[U(t)(\rho_s(0) \otimes \rho_b(0))U(t)^*]$$

and we can easily derive a qsde for it. A system Heisenberg observable X after time t evolves to

$$X(t) = j_t(X) = U(t)^*XU(t) = U(t)^*(X \otimes I)U(t)$$

and its evolution is of the form

$$dj_t(X) = j_t(\theta_0(X))dt + \sum_k j_t(\theta_k(X))dB_k(t), W_k(t) = B'_k(t)$$

and then

$$\begin{aligned} dTr(\rho_s(t)X) = dTr(\rho(t)X) = dTr(\rho(0)j_t(X)) = \\ Tr(\rho(0)dj_t(X)) = Tr(\rho(0)j_t(\theta_0(X))dt + Tr(\rho(0)(j_t(\theta_k(X))dB_k(t))) \\ = Tr(\rho(t)\theta_0(X))dt + Tr(U(t)dB_k(t)\rho(0)U(t)^*\theta_k(X)) \\ = Tr(\rho_s(t)\theta_0(X))dt + Tr(U(t)dB_k(t)U(t)^*\rho(t)\theta_k(X)) \end{aligned}$$

Note that $dB_k(t)$ acts in $\Gamma_s(\mathcal{H}_{[t+dt,\infty)})$ while $U(t)$ acts in $\mathfrak{h} \otimes \Gamma_s(\mathcal{H}_{t]})$ and $\theta_k(X)$ acts in $\mathfrak{h}$. Therefore, this equation can also be expressed as

$$Tr(d\rho_s(t)X) = Tr(\theta_0^*(\rho_s(t))X)dt + Tr(dB_k(t)\rho(t)\theta_k(X))$$

Note that although $U(t)$ acts in $\Gamma_s(\mathcal{H}_{t]})$, $\rho(t)$ acts in the whole of $\mathfrak{h} \otimes \Gamma_s(\mathcal{H})$ and not just in $\Gamma_s(\mathcal{H}_{t]})$ because $\rho(0)$ acts in the whole of $\mathfrak{h} \otimes \Gamma_s(\mathcal{H})$ in general which can be seen for example by considering the example when $\rho_b(0)$ is a coherent state of the bath:$\rho_b(0) = |\phi(u)><\phi(u)|$ with $|\phi(u)>= exp(-\parallel u \parallel^2/2)|e(u)>$ for some $u \in \mathcal{H}$. The differential equation satisfied by $\rho_s(t)$ is therefore not a closed form equation, ie, it also involves the bath state at time t and in fact the entire system and bath state at time t:

$$d\rho_s(t) = \theta_0^*(\rho_s(t))dt + Tr_2((\theta_k \otimes I)^*(dB_k(t)\rho(t))$$

Suppose however that the bath is initially the coherent state $|\phi(u)>$. Then we easily deduce the result that

$$Tr(dB_k(t)\rho(t)\theta_k(X)) = f_k(u(t), \bar{u}(t))dt.\theta_k^*(\rho_s(t))$$

for some function $f_k(z, \bar{z})$ of the complex vector z and its conjugate. Then,

$$d\rho_s(t)/dt = \theta_0^*(\rho_s(t)) + f_k(u(t), \bar{u}(t))\theta_k^*(\rho_s(t))$$

the summation over k being understood. We can now by further partial tracing of this equation over the other particles, derive an open quantum system approximate Boltzmann equation for the marginal state of each particle assuming all the particles to be indistinguishable: The exact partially traced equation is

$$d\rho_{s1}(t)/dt = Tr_{23...N}(\theta_0^*(\rho_s(t)) + f_k(u(t), \bar{u}(t))Tr_{23...N}(\theta_k^*(\rho_s(t)))$$

We assume that

$$\theta_0(X) = i[H, X] - (1/2) \otimes_{k=1}^N \theta_{0k}$$

where

$$H = \sum_k H_k + \sum_{k<j} V_{kj}$$

with H_k acting in the Hilbert space $\mathfrak{h}_k$ of the kth particle, V_{kj} acting in the tensor product space $\mathfrak{h}_k \otimes \mathfrak{h}_j, k < j$ and θ_{0k} acting the algebra $\mathcal{B}(\mathfrak{h}_k)$ of operators defined in $\mathfrak{h}_k$.

3.5 The pde's satisfied by the quantum electromagnetic field observables in a cavity resonator in the presence of bath noise

$$H_s = \sum_{k=1}^{N} \omega(k) a(k)^* a(k)$$

is the cavity field Hamiltonian. The cavity vector potential field has the form

$$A(t,r) = \sum_k (a(k)\phi_k(t,r) + a(k)^* \phi_k(t,r)^*)$$

where $\phi_k(t,r)$ are complex valued functions of time and space. Here,

$$[a(k), a(m)^*] = \delta(k,m), [a(k), a(m)] = [a(k)^*, a(m)^*] = 0$$

We consider perturbations to the cavity Hamiltonian of the form

$$\Delta H_s = \epsilon.f(a(k), a(k)^*, k = 1, 2, ..., N)$$

and derive a nonlinear pde with perhaps integro-differential terms satisfied by $A(t,r)$. Note that

$$[a(k), f(a,a^*)] = f_{,2k}(a,a^*), [a(k)^*, f(a,a^*)] = -f_{,1k}(a,a^*)$$

and note that assuming spatial orthogonality of the functions ϕ_k:

$$\int A(t,r)\phi_k^*(t,r)d^3r = a(k), \int A(t,r)\phi_k(t,r)d^3r = a(k)^*$$

These identities can be used to derive integro-differential equations satisfied by $A(t,r)$. Thus, we can write

$$\partial_t A(t,r) = i[H_s + \Delta H_s, A(t,r)] = \sum_k (-i\omega(k)\phi_k(t,r)a(k) + i\omega(k)\phi_k(t,r)^* a(k)^*)$$

$$+\epsilon \sum_k [if_{,2k}(a,a^*)\phi_k(t,r) - if_{,1k}(a,a^*)\phi_k(t,r)^*]$$

Also,

$$f_{,2k}(a,a^*) = f_{,2k}(\int A(t,r)\phi(t,r)^* d^3r, \int A(t,r)\phi(t,r)d^3r)$$

$$f_{,1k}(a,a^*) = f_{,1k}(\int A(t,r)\phi(t,r)^* d^3r, \int A(t,r)\phi(t,r)d^3r)$$

Also,

$$\partial_t^2 A(t,r) = i[H_s + \Delta H_s, \partial_t A(t,r)]$$

References:

[1] Timothy Eyre, "Quantum stochastic calculus and representations of Lie super-algebras", Springer lecture notes.

[2] Lec Bouten, "Filtering and control in quantum optics", Ph.D thesis.

3.6 Estimating the quantum state of a single particle in a system of N indistinguishable particles interacting with each other and with an external bath field

The marginal density of (one particle) after partial tracing out over the other particles and over the bath satisfies approximately the quantum Boltzmann equation

$$i\partial_t\rho_1(t) = [H_1, \rho_1(t)] + (N-1)Tr_{23...N}[V_{12}, \rho_1(t) \otimes \rho_1(t)]$$

$$+\theta_1(\rho_1(t))$$

Note that the partial trace of the Lindblad operator assumed to be separable, acting on the state is given by

$$Tr_{23...N}(L_1^{\otimes N} L_1^{*\otimes N}\rho(t) + \rho(t)L_1^{\otimes N}L_1^{*\otimes N} - 2L_1^{*\otimes N}\rho(t)L_1^{\otimes N})$$

If we approximate $\rho(t)$ by $\rho_1(t)^{\otimes N}$ in the Lindblad term, (This is justified since the Lindlbad operators being small, the contributions of the first order of smallness terms in $\rho(t)$ to the Lindblad terms will be of the second order of smallness. Thus, we get approximately,

$$Tr_{23...N}(L_1^{\otimes N} L_1^{*\otimes N}\rho(t) + \rho(t)L_1^{\otimes N}L_1^{*\otimes N} - 2L_1^{*\otimes N}\rho(t)L_1^{\otimes N})$$

$$= Tr(L_1L_1^*\rho_1(t))^{N-1}[L_1L_1^*\rho_1`(t) + \rho_1(t)L_1L_1^* - 2L_1^*\rho_1(t)L_1]$$

Thus, our quantum Boltzmann equation for separable open quantum systems assumes the form

$$i\partial_t\rho_1(t) = [H_1, \rho_1(t)] + (N-1)Tr_{23...N}[V_{12}, \rho_1(t) \otimes \rho_1(t)]$$

$$-(1/2)(Tr(L_1L_1^*\rho_1(t))^{N-1}[L_1L_1^*\rho_1`(t) + \rho_1(t)L_1L_1^* - 2L_1^*\rho_1(t)L_1])$$

This is a nonlinear operator differential equation for $\rho_1(t)$.

Remark: Suppose for example we take $L_1 = aq_1 + bp_1$ where (q_1, p_1) is the canonical position-momentum pair for the first particle. Then, in the position representation, $\rho_1(t)$ is represented by the continuous matrix $< q_1'|\rho_1(t)|q_1 >= \rho_1(t, q_1', q_1)$ and we get

$$[L_1^* \rho_1)](t, q_1, q_1') =$$

Chapter 4

LDP Applied to Quantum Measurement, Classical Markov Chains, Quantum Stochastics and Quantum Transition Probabilities

4.1 Some other aspects of measurement of a quantum field

Let $X(r), r \in D$ be a quantum field and let $X_k = X(r_k), k = 1, 2, ..., N$ be the spatially discretized quantum field. Let

$$X_k = \sum_r |e_{k,r} > \lambda(k,r) < e_{k,r}|$$

be the spectral decompostion of X_k. Let $\rho(\theta)$ be the initial state of the entire quantum field. Let $H(t)$ denote the Hamiltonian of the quantum field. Let $U(t,s) = T(exp(-i\int_s^t H(u)du)), t > s$ denote the evolution operator. We measure X_k at time t_k- causing the state $\rho(\theta, t_k-)$ to collapse to the state $\rho(\theta, t_k+) = \sum_r M_{k,r}\rho(\theta, t_k-)M_{k,r}$ where $M_{k,r} = |e_{k,r} >< e_{k,r}|$. Then this state evolves under the Hamiltonian $H(.)$ to give the state $\rho(\theta, t_{k+1}-)$

$$= U(t_{k+1}, t_k)\rho(\theta, t_k+)U(t_{k+1}, t_k)^*$$

at time $t_{k+1}-$. After this measurement of X_k at time t_k, if the measurement outcome is not noted, the output of the measuring apparatus is given by $Tr(\rho(\theta, t_k-)X_k)$. In this way, the total sequence of the measurement outputs is given by $Tr(\rho(\theta, t_k-)X_k), k = 1, 2, ..., N$ and from these outputs, we can estimate the parameter θ on which the initial state of the system depends. If the quantum field and the state are objects of an open quantum system, then the unitary evolution operator $U(t,s)$ should be replaced by the quantum dynamical semigroup $T(t,s)^*$ obtained by solving the Lindblad equation, ie, the master equation. More precisely, the state

$$\rho(\theta, t_{k+1}-) = T(t_{k+1}, t_k)^*(\rho(\theta, t_k+))$$

where $T(t,s)$ solves the master equation:

$$\partial T(t,s)(X)/\partial t = \theta_t(T(t,s)(X)), t \geq s, T(s,s) = I$$

with $\theta_t(.)$ being the Lindblad operator that generates the quantum dynamical semigroup:

$$\theta_t(X) = i[H, X] - (1/2)\sum_k (L_k^* L_k X + X L_k^* L_k - 2L_k^* X L_k)$$

Note that this formalism of state collapse during measurements followed by a free evolution between two measurements is in agreement with the Schrodinger picture in which the states evolve both due to the noisy Hamiltonian and due to measurements while the quantum Heisenberg fields do not evolve in time. It is not possible to give a Heisenberg picture of measurements since it is meaningless to talk about observable collapse during a measurement. For example, under Heisenberg dynamics, the observable at time t_1 is $T(t_1, 0)(X_1)$. Its average value at this time is $Tr(\rho(\theta)T(t_1, 0)(X_1)) = Tr(T(t_1, 0)^*(\rho(\theta))X_1)$ which agrees with the Schrodinger picture. However after the measurement of X_1 at time t_1, only the state collapses not the observable. In fact, following this measurement, the state at time t_1+ becomes $\rho(\theta, t_1+) = M_1(\rho(\theta, t_1-)) = \sum_r M_{1,r}\rho(\theta, t_1-)M_{1,r}$. The average of X_1 in this state in accord to the Schrodinger picture is given by

$$Tr(\rho(\theta, t_1+)X_1) = Tr(M_1(\rho(\theta, t_1-))X_1) =$$

$$Tr(\rho(\theta, t_1-)M_1^*(X_1)) = Tr(T(t_1, 0)^*(\rho(\theta))M_1^*(X_1))$$
$$= Tr(\rho(\theta)T(t_1, 0)(M_1^*(X_1))$$

so we can in fact give meaning to observable evolution based on free evolution followed by measurement by the equation

$$X_1 \to T(t_1, 0)(M_1^*(X_1)) = (M_1 o T(t_1, 0)^*)^*(X_1)$$

More generally, the Heisenberg evolution of the system from time 0 to time t_N taking measurements at times $t_1 < t_2 < ... < t_N$ into account is given by the dual of the Schrodinger evolution:

$$(M_N T(t_N, t_{N-1})^* ... M_2 T(t_2, t_1)^* M_1 T(t_1, 0)^*)^* =$$

$$T(t_1, 0)M_1^* T(t_2, t_1)M_2^* ... T(t_N, t_{N-1})M_N^*$$

In other words, the dual Heisenberg map reverses the Schrodinger map. This is a non-causal operation ie future terms appear first and the past terms appear ater. This is why it is more convenient to handle measurements and state collapse more easily using the Causal Schrodinger picture. It should be noted that the Schrodinger evolution map from time t_k- to time $t_{k+1}-$ is given by

$$T(t_{k+1}, t_k)^*.M_k$$

and the Schrodinger map from time t_k+ to time $t_{k+1}+$ is given by

$$M_{k+1}T(t_{k+1}, t_k)^*$$

4.2 Some parts of the solution to the question paper on antenna theory

2a] $L = 0.075\lambda$. $I(z) = I_0 sin(\beta(L/2 - |z|))$, $\beta = 2\pi/\lambda$, $\beta L/2 = 0.24$. The spatial wave number of the current is $0.24 << \pi/2$. So over the range $[0, 0.24]$, the sine wave is effectively a triangle. Equivalently

$$I(z) = I_0 sin(0.24(1 - 2|z|/L)) \approx I_0.2.4.(1 - 2|z|/L), |z| \leq L/2$$

Here we have used the approximation

$$sin(x)/x \approx 1, |x| << \pi/2$$

Antenna pattern: Frequency $\omega, K = \omega/c = 2\pi/\lambda$.

$$A_z(r,\theta) = (\mu/4\pi r)exp(-jKr)\int_{-L/2}^{L/2} I(z)exp(jKzcos(\theta))dz$$

$$\approx (\mu/4\pi r)exp(-jKr)0.24.\int_{-L/2}^{L/2}(1 - 2|z|/L)exp(jKzcos(\theta))dz$$

Note that

$$\int_{-L/2}^{L/2}(1-2|z|/L)exp(jKz.cos(\theta))dz = \lambda\int_{0.75/2}^{0.75/2}(1-2\xi/0.75)exp(j2\pi\xi.cos(\theta))d\xi$$

$$= 2\lambda.\int_0^{0.75/2}(1 - 2\xi/0.75)exp(j2\pi\xi.cos(\theta))d\xi$$

The electric and magnetic fields in the far field zone are computed from the above vector potential $A = A_z\hat{z}$ as follows:

$$B = curlA, curlcurlA = curlB = j\omega\epsilon\mu E$$

$$E = curlcurlA/j\omega\epsilon\mu = (j\omega\epsilon\mu)^{-1}(\nabla(divA) - \nabla^2 A)$$

$$= (j\omega\epsilon\mu)^{-1}(\nabla(divA) + K^2A)$$

This gives

$$E_\theta = (-K^2/j\omega\epsilon\mu)A_z.sin(\theta) + O(1/r^2)$$

$$= (-K^2I_.0.24/(j.4\pi\omega\epsilon r))sin(\theta).2\int_0^{L/2} cos(Kz.cos(\theta)).(1-2|z|/L)dz$$

Total power radiated by the linear antenna:

$$P = 2\pi r^2\int_0^\pi (|E_\theta|^2/2\eta)sin(\theta)d\theta, \eta = \sqrt{\mu/\epsilon}$$

$$= (\pi/\eta)\int_0^\pi sin^3(\theta)(K^2I_0.0.24/4\pi\epsilon\omega)^2.4(\int_0^{L/2} cos(Kz.cos(\theta))(1-2z/L)dz)^2$$

$$= I_0^2R_r/2$$

Thus the radiation resistance of the linear antenna is given by

$$R_r = (\pi/\eta)K^40.24^2.8/(4\pi\epsilon\omega)^2\int_0^\pi sin^2(\theta)d\theta.(\int_0^{L/2} cos(Kz.cos(\theta))(1-2z/L)dz$$

[b] $L = 1.5\lambda$.

$$I(z) = I_0.sin((2\pi/\lambda)(1.5\lambda/2-|z|) = I_0.sin(1.5\pi(1-2|z|/1.5\lambda)), |z| \leq L/2 = 0.7$$

or equivalently,

$$I(z) = I_0.sin(1.5\pi(1-|\xi|), |\xi| \leq 1$$

The current pattern consists of three half sine-wave lobes from $\xi = -1$, to $\xi = 1$

[2] Solve by the method of images. The positions of the centres of the infinitesimal antenna dipoles are $((n+1/2)a, mb, h), h = \lambda/4, n = -2, -1, 0, 1, m = -2, -1, 0, 1, 2$ where $a = b = \lambda/4$ are the interelement spacings along the x and y directions. The positions of the image antennas are $(n + 1/2a, mb, -h)), n = -2, -1, 0, 1, m = -2, -1, 0, 1, 2$. The orientation of the image dipoles are reversed since a positive charge has a negative charge as its image and vice versa. The generated magnetic vector potential in the far field zone is

$$A = A_z\hat{z} = (exp(-jKr)/r)(\mu I_0\hat{z}/4\pi)$$

$$\sum_{-2\leq n\leq 1, -2\leq m\leq 2} exp(jK((n+1/2)a.cos(\phi)sin(\theta)+mb.sin(\phi)sin(\theta)+h.cos(\theta))$$

$$-(exp(-jKr)/r)(\mu I_0\hat{z}/4\pi)$$

$$\sum_{-2\leq n\leq 1, -2\leq m\leq 2} exp(jK((n+1/2)a.cos(\phi)sin(\theta)+mb.sin(\phi)sin(\theta)-h.cos(\theta)))$$

which sums up to give

$$A_z = (\mu I_0exp(-jKr)/4\pi r)(4j)(cos(\pi.cos(\phi)cos(\theta)/2)+cos(3\pi.cos(\phi)sin(\theta)/2)$$

$$.(1+2.cos(\pi.sin(\phi)sin(\theta))+2.cos(2\pi.sin(\phi)sin(\theta)))$$
$$=(j\mu I_0.exp(-jKr)/\pi r)P(\theta,\phi)$$

where

$$P(\theta,\phi)=(cos(\pi.cos(\phi)cos(\theta)/2)+cos(3\pi.cos(\phi)sin(\theta)/2)).$$
$$(1+2.cos(\pi.sin(\phi)sin(\theta)+2.cos(2\pi.sin(\phi)sin(\theta))$$

The far field electric field pattern is

$$E_\theta=-j\omega A_z.sin(\theta)=(\omega\mu exp(-jKr)/\pi r)P(\theta,\phi)sin(\theta)$$

[3] Use the formula for the resonant frequencies within a rdra (rectangular dielectric resonator antenna)

$$\omega(mnp)=(\pi/\sqrt{\epsilon\mu})(m^2/a^2+n^2/b^2+p^2/d^2)^{1/2},m,n,p=0,1,...$$

to do the design, ie, select a,b,d for given resonant frequencies. The field within the rdra is assumed to be TM so that it can be derived from A_z, the only non-zero component of the magnetic vector potential:

$$(\nabla^2+K^2)A_z(x,y,z)=0,K=\omega\sqrt{\epsilon\mu},H=curlA/\mu$$

$$curlcurlA=j\omega\epsilon E$$

or

$$j\omega\epsilon E=\nabla(\nabla.A)+K^2A$$

Thus,

$$j\omega\epsilon E_z=\partial^2A_z/\partial z^2+K^2A_z$$
$$j\omega\epsilon E_x=\partial^2A_z/\partial x\partial z,$$
$$j\omega\epsilon E_y=\partial^2A_z/\partial y\partial z$$

E_z must vanish when $x=0,a,y=0,b$ and E_x,E_y must vanish when $z=0,d$. Therefore, the appropriate solution is

$$A_z=C.sin(m\pi x/a)sin(n\pi y/b)cos(p\pi z/d)$$

The radiation pattern produced by the sidewalls and the top surface of this rdra can be found out by calculating the tangential components of the magnetic field on these surfaces and hence the surface current densities and then by applying the standard far field approximation formulas for the retarded potentials generated by these surface current densities. Once the radiation pattern has been determined in the far-field zone, we can evaluate the Poynting vector upto $O(1/r^2)$ terms and then integrate over all solid angles to obtain the radiation resistance of the rdra as a function of a,b,d.

When the antenna is connected to a 50Ω microstrip line then matching must be achieved, ie, the radiation resistance of the antenna must be equal to the characteristic impedance 50Ω of the feeding microstrip line. This matching condition gives us constraints on a, b, d which facilitate the design process.

[4] According to Maxwell's equations the equation of wave propagation in a conducting medium is derived from

$$curlE = -j\omega\mu H, curlH = (\sigma + j\omega\epsilon)E, divE = 0, divH = 0$$

Thus,

$$curlcurlE = -j\omega curlH = -j\omega\mu(\sigma + j\omega\epsilon)E$$

or equivalently,

$$(\nabla^2 - j\omega\mu(\sigma + j\omega\epsilon))E = 0$$

The imaginary part of the propagation constant $\gamma = \sqrt{j\omega\mu(\sigma + j\omega\epsilon)}$ is the wave number of propagation. Denoting it by β, the wavelength of the propagating wave is $\lambda = 2\pi/\beta$ and the phase velocity can be determined as $v_p = \omega/\beta$ while the group velocity is $v_g = (d\beta/d\omega)^{-1}$. The refractive index of the medium is c/v_g and we denote it by $n(\omega, x, y, z)$ for a fixed frequency. Here, we are assuming ϵ, μ to be slowly varying functions of ω, x, y, z. Neglecting attenuation, the time taken for a light wave to propagate from point 1 from point 2 is given by

$$T(1,2) = \int_1^2 n(x(s), y(s), z(s))ds/c$$

and the effective transmission path length of the ray is therefore

$$cT(1,2) = \int_1^2 nds$$

[5] Loop antenna used for direction finding. Assume that the loop is of radius R and its plane is in the xy plane with centre at the origin. Then the only non-vanishing component of the far field magnetic vector potential is

$$A_\phi(r, \theta, \phi) = (\mu I.exp(-jKr)/4\pi r) \int_0^{2\pi} exp(jKRsin(\theta)cos(\phi'))cos(\phi')d\phi'$$

$$= (-j\mu I/2r)exp(-jKr)J_0'(KR.sin(\theta))$$

where

$$J_0(x) = (2\pi)^{-1} \int_0^{2\pi} exp(jx.cos(\phi))d\phi$$

The maximum of A_ϕ occurs at $\theta = 0$ and this directivity is large:

$$D = \frac{2|J_0'(0)|^2}{\int_0^\pi |J_0(KR.sin(\theta))|^2 sin(\theta)d\theta}$$

The far field electric field pattern is given by

$$E_\phi = -j\omega A_\phi$$

The corresponding Poynting vector in the far field zone is

$$P = |E_\phi|^2 \hat{r}/2\eta$$

Due to the large directivity of the loop antenna, it can be used for accurate direction finding. If the antenna is pointed towards the given target with large direction error, due to its large directivity and hence small relative sidelobe level, it will not collect the required signal. It will collect the required signal only when the direction error is very small since large directivity implies a very narrow and large mainlobe relative to the sidelobe.

If V is the emf induced in the loop and E_x is the field strength assumed to be constant and along the x axis, then

$$V = -\int_0^{2\pi} E_x sin(\phi) d\phi = 0$$

However, if the electric field varies with position, we write

$$E = E_x(x,y)\hat{x} + E_y(x,y)\hat{y}$$

the emf induced by this field in the circular loop is then given by

$$V = \int_0^{2\pi} (-E_x(R.cos(\phi), R.sin(\phi)) + E_y(R.cos(\phi), R.sin(\phi)))Rd\phi$$

and from this equation, the field strength of the signal picked may be determined.

[6] [a] The positions of the sensors in space are given by

$$(R.cos(2k\pi/8)\hat{x} + R.sin(2\pi k/8)\hat{y}), k = 0, 1, ..., 7$$

and hence the array factor is given by

$$A(\theta,\phi) = I_0 \sum_{k=0}^{7} [exp(jkR.cos(2k\pi/8)cos(\phi)sin(\theta) + jkR.sin(2k\pi/8)sin(\phi)sin(\theta))$$

$$= I_0 \sum_{k=0}^{7} exp(jkR.sin(\theta).cos(\phi - 2k\pi/8))$$

This formula justifies the fact that the array factor can be found from the concept of planar array factor.

[b] If the electric field is clockwise circularly polarized, then it can be expressed as

$$E(t,\theta,\phi) = E_0(\theta,\phi)(\hat{x}.cos(\omega t) - \hat{y}.sin(\omega t))$$

and hence the time averaged power collected by the sensor is given by

$$< (E_0(\theta,\phi)f(r,\theta,\phi))^2 < (2.cos(\omega t) - sin(\omega t))/\sqrt{5})^2 >$$

$$= (E_0(\theta,\phi)f(r,\theta,\phi))^2/2 = L(\theta,\phi)$$

say. Here, we have used

$$< cos^2(\omega t) >=< sin^2(\omega t) >= 1/2, < cos(\omega t).sin(\omega t) >= 0$$

Thus, the loss factor is

$$E_0(\theta,\phi)^2 - L(\theta,\phi) = 1 - f(r,\theta,\phi)^2/2$$

4.3 Some additional LDP related problems in antenna theory

[1] Excite an atom having an electron or a set of N electrons by an electromagnetic field produced by an antenna driven by a sinusoidal current source plus a weak Gaussian plus Poisson noise. Compute the transition probability of the electrons making a transition between two stationary states under the influence of this radiation using first or even second order perturbation theory. Then, calculate the rate function for these transition probabilities and hence evaluate the approximate probability using this rate function that the electron's transition probability will deviate from the case of noisless current by an amount greater than a threshold. In practise, while designing a laser, we are interested in having a desired population of electrons in a given excited state so that when the source of radiation is removed, the electrons make a transition to the ground state thereby emitting monochromatic radiation with the desired intensity which is proportional to the population of electrons in the excited state which is in turn proportional to the transition probability of the electrons to the excited state. So our LDP approach gives us the small probability of the laser so designed to fail to produce the given frequency and intensity of radiation.

[2] Consider a straight wire antenna carrying a current $I_0 sin(\beta(L/2 - |z|))$. If L, I_0 undergo small random fluctuations, then find out the rate function of the generated electromagnetic field in the far field zone and hence calculate the approximate probability that the pattern will deviate from the noiseless pattern over a finite region of space-time by an amount greater than a given threshold ϵ. Now design an additive non-random control current source $I_c(z|\theta)$ by choosing the parameters θ appropriately so that this deviation probability after taking into account the control current is a minimum. Work out this problem using [a] Approximate perturbation theory and [b] the exact contraction principle in large deviation theory.

[3] The far field power flux pattern of an antenna is $P(\theta,\phi)\hat{r}/r^2$. Our sensor can measure the total power $\int_{\Omega_0} P(\theta,\phi)sin(\theta)d\theta.d\phi$ passing through the solid angle Ω_0. If the solid angle Ω_0 of our aperture sensor flucuates from $\Omega_0 = [\theta_0, \theta_0+\Delta\theta_0] \times [\phi_0, \phi_0+\Delta\phi_0]$ to $\Omega_0 + \delta\Omega_0$ where $\Delta\Omega_0$ is a random solid angle, then calculate the rate function of the fluctuation

$$\int_{\Omega_0+\Delta\Omega_0} P(\hat{r})d\Omega(\hat{r}) - \int_{\Omega_0} P(\hat{r})d\Omega(\hat{r})$$

given the statistics of the indicator random field $\chi_{\Omega_0+\Delta\Omega_0}(\hat{r})$.

[4] Let $\mathbf{E}(t,x,y,z)$ be an electromagnetic field. Assume that this field is random and stationary in space-time. Let $\mathbf{l}(t)$ be an antenna sensor located at x, y, z and with direction varying with time. The induced emf in this sensor is given by

$$V(t,x,y,z) = (\mathbf{E}(t,x,y,z), \mathbf{l}(t))$$

If this sensor moves slowly along a trajectory $(x(t), y(t), z(t))$ over the time duration $[0,T]$, then the time average emf induced in it is given by

$$V_0(T) = T^{-1}\int_0^T V(t,x(t),y(t),z(t))dt$$

Assume that $\mathbf{l}(t)$ is also a stationary process and so is the trajectory process $(x(t), y(t), z(t))$ with all the three processes being independent of each other. Then, calculate the LDP rate function for $V_0(T), T \to \infty$. For this, specifically, evaluate the limiting logarithmic moment generating function

$$\Lambda(\lambda) = lim_{T\to\infty}T^{-1}.log(\mathbb{E}(exp(TV_0(T)))$$

[5] The magnetic vector potential satisfies the three dimensional wave equation with white Gaussian noise current density source. Its pde can be written as

$$dA_\epsilon(t,r) = V_\epsilon(t,r), dV_\epsilon(t,r) = \nabla^2 A_\epsilon(t,r) + \sqrt{\epsilon}J(t,r)$$

where ϵ is a small parameter and $J(t,r)$ is white w.r.t the time variable, ie,

$$\mathbb{E}(J(t,r)J(s,r')^T) = \delta(t-s)R_{JJ}(r,r')$$

write this down as an Ito stochastic differential equation w.r.t the time variable by writing

$$\int_0^t J(s,r)ds = \sum_k B_k(t)\phi_k(r)$$

with the $B_k's$ being standard independent Brownian motion processes and

$$R_{JJ}(r,r') = \sum_k \phi_k(r)\phi_k(r')^T$$

being the Karhunen-Loeve expansion of the positive definite kernel R_{JJ}. The by application of Schlider's theorem, derive the rate functional for the Gauss Markov process field $(A_\epsilon(t,r), V_\epsilon(t,r) : r \in \mathbb{R}^3)$ as $\epsilon \to 0$. Now set $\epsilon = 1$ and compute the rate function of the empirical density functional of the Marko process $(A(t,r), V(t,r)) : r \in \mathbb{R}^3$. For doing this, you must first evaluate the transition probability density generator kernel of this process by using Ito' formula for Brownian motion Specifically, note that the system of stochasti pde's can be reduced to an infinite sequence of sde's for the processes

$$A_k(t) = \int (A(t,r), \phi_k(r))d^3r, V_k(t) = \int (V(t,r), \phi_k(r))d^3r, k = 1, 2, ...$$

driven by the infinite sequence of Brownian motion processes $B_k(t), k = 1, 2, ...$ Make use of the identities

$$A(t,r) = \sum_k A_k(t)\phi_k(r), V(t,r) = \sum_k V_k(t)\phi_k(r)$$

and hence the relations

$$dA_k(t) = dt \int (V(t,r), \phi_k(r))d^3r = V_k(t)dt,$$

$$dV_k(t) = dt \int (\nabla^2 A(t,r) + J(t,r), \phi_k(r))d^3r = dt \sum_m A_m(t) < \nabla^2\phi_m, \phi_k >$$
$$+dB_k(t)$$

where

$$< \nabla^2\phi_m, \phi_k >= \int \phi_k(r)\nabla^2\phi_m(r)d^3r$$

4.4 LDP for quantum Markov chains using discrete time quantum stochastic flows

Let θ^a_b denote the structure maps. The discrete time Markovian quantum stochastic flow is given by

$$j_{n+1}(X) = \sum_{a,b} j_n(\theta^a_b(X)) \otimes |e_a >< e_b| \otimes I_{[n+2,\infty)}$$

where $\{|e_a >: 0 \leq a \leq N\}$ is an onb for the Hilbert space $\mathcal{H}$ in which the observable X is defined. Thus assuming that j_n is a $*$ unital homomorphism from $\mathcal{B}(\mathcal{H})$ into $\mathcal{B}(\mathcal{H}_{n]})$ (Note that $\mathcal{H}_{0]} = \mathcal{H}$), it follows that

$$j_{n+1}(X)j_{n+1}(Y) = \sum_{a,b,c,d} j_n(\theta^a_b(X)\theta^c_d(Y)) \otimes |e_a >< e_b|e_c >< e_d > \otimes I_{[n+2,\infty)}$$

$$= \sum_{a,b,d} j_n(\theta_b^a(X)\theta_d^b(Y)) \otimes |e_a > e_d| \otimes I_{[n+2,\infty)}$$

$$= \sum_{a,d} j_n(\theta_d^a(XY)) \otimes |e_a >< e_d| \otimes I_{n+2,\infty)} = j_{n+1}(XY)$$

provided that we assume the structure equations

$$\theta_b^a(XY) = \sum_c \theta_c^a(X)\theta_d^c(Y)$$

Thus, by induction, it follows that $j_n(.)$ is a star unital homomorphism for each $n \geq 0$ provided that we take $j_0(X) = X$. Note that

$$\mathcal{H}_{n]} = \mathcal{H}^{\otimes n}, n \geq 0$$

and

$$\mathcal{H}_{[n,\infty)} = \otimes_{n \geq k} \mathcal{H}$$

Now we can choose a state say $|\phi >= |e_0 > \otimes |e_0 > \otimes ...$ for the system and evaluate the conditional expectation

$$\mathbb{E}_{n]} j_{n+1}(X) = \mathbb{E}_{\phi_{[n+1,\infty)}}[j_{n+1}(X)]$$

where

$$\phi_{[n,\infty)} = \otimes_{[n+1,\infty)} |e_0 >$$

is the state on $\mathcal{H}_{n+1,\infty)}$. This conditional expectation is easily seen to evaluate to

$$j_n(T(X)), T = \theta_0^0$$

θ_0^0 is called the generator of the quantum Markov process $\{j_n(.) : n \geq 0\}$ Now we wish to ask the following question. For large N, calculate the probability distribution of the observable

$$Z_N = N^{-1}.\sum_{n=1}^{N} j_n(X)$$

For this, we need to evaluate first the logarithmic moment generating function of NZ_N. It is given by

$$\Lambda_N(NF) = log\mathbb{E}_\phi exp(N\,Tr(FZ_N))$$

Then, we need to evaluate the limiting logarithmic moment generating function

$$\bar{\Lambda}(F) = lim_{N\to\infty} N^{-1}.\Lambda_N(F)$$

and finally the Legendre transform of this:

$$\Lambda^*(Z) = sup_F(Tr(FZ) - \bar{\Lambda}(F))$$

The first step in evaluating the moment generating function of $N.Z_N$ is to evaluate $\mathbb{E}_\phi(exp(A+B))$ for two non-commuting observables in terms of the moments

$$\mathbb{E}_\phi(A^{m_1}B^{m_2}A^{m_3}B^{m_4}...A^{m_{2n-1}}B^{m_{2n}})$$

4.5 LDP applied to the analysis of the error process in stochastic filtering theory of a continuous time Markov process when the measurement noise is white Gaussian and more generally when the measurement noise is the differential of a Levy process (ie, a limit of compound Poisson process plus white Gaussian noise)

.

Statement of the problem: The state of the system satisfies a vector valued sde driven by weak vector valued standard Brownian motion. The measurement process at any given time is a function of the state plus measurement noise. The desired trajectory satisfies the same differential equation as the the state process but without process noise. The EKF is formulated which gives a sde satisfied by the state estimate driven by the measurement process. The difference between the desired state trajectory and its EKF estimate is fed-back as a force term in the state process sde. More generally, we could include a disturbance in the state equations and replace this disturbance by the disturbance observer output plus white noise. In this case, the disturbance observer output becomes a part of the set of state variables. From these differential equations, derive approximate linearized stochastic differential equations for the EKF state estimation error and the trajectory tracking errors driven by linear combinations of the state process noise, the white noise coming from the disturbance estimate error and the measurement noise. We then solve these linear sde's over a given time duration and calculate the rate functionals of this total error process, ie, the state estimation and trajectory tracking errors. This rate functional is used to calculate the approximate probability for the error to deviate from zero by an amount greater than a given threshold and the parameters of the system including the Gain matrix in the feedback term are adjusted for this deviation probability to be minimized.

4.6 LDP applied to the electroweak theory

ϕ is the scalar Gibbs field. ψ is the electron-Lepton field. A_μ is the Gauge boson field comprising the W and Z bosons which acquire mass after symmetry breaking and the photon field which does not acquire mass after symmetry breaking. The total Lagrangian is

$$L(\psi, A, phi) = (-1/4)Tr(F_{\mu\nu}F^{\mu\nu}) + \bar{\psi}(\gamma.(i\partial + eA))\psi$$

$$+(1/2)(|(\partial + ieA)\phi|^2 - m_H^2\phi^2)) + Re[g^*(\psi \otimes \bar{\psi} \otimes \phi)]$$

Note the shorthand notations used:

$$|(\partial + ieA)\phi|^2 = (\partial_\mu + ieA_\mu)\phi)^*(\partial^\mu - ieA^\mu)\phi,$$

$$\gamma.(i\partial + eA) = \gamma^\mu(i\partial_\mu + eA_\mu)$$

In the absence of any interactions between the gauge fields, the electron-lepton fields and the Higgs field, the gauge fields satisfy the free Yang-Mills field equations

$$D_\nu F^{\mu\nu} = 0$$

which is a short-hand notation for

$$\partial_\nu F^{\mu\nu a} + ieC(abc)A^b_\nu F^{\mu\nu c} = 0$$

where

$$F^{\mu\nu a} = \partial A^{\nu a,\mu} - A^{\mu b,\nu} + eC(abc)A^{\mu b}A^{\nu c}$$

is the gauge field tensor, the Higgs field satisfies the free Klein Cordon equation

$$(\partial_\mu\partial^\mu + m_H^2)\phi = 0,$$

and the electron-Lepton field satisfies the free Dirac equation with zero mass.

$$i\gamma^\mu\partial_\mu\psi = 0$$

The solutions to these free field equations (called the unperturbed equations) by setting in addition, the nonlinear terms in the Yang Mills equations to zero, are given by plane wave expansions, with the coefficients in the plane wave expansion of the Yang-Mills field being the Gauge boson creation and annihilation operators in momentum-helicity space, the coefficients in the plane wave expansion of the Higgs field being the Higgs boson creation and annihilation operators in momentum space and the coefficients in the plane wave expansion of the electron-Lepton field being the electron annihilation and positron creation operator fields in momentum-spin space. These zeroth order approximate solutions we denote by $A_\mu|^{(0)}, \phi^{(0)}$ and $\psi^{(0)}$. We then apply perturbation theory to the exact equations taking into account nonlinearities and interactions (regarded as being of the first order of smallness) and then obtain perturbation expansions for the perturbed fields which are given by quadratic, cubic and higher order combinations of the unperturbed creation and annihilation operators. Having fixed our order of perturbation, we substitute these into the Hamiltonian of the system to express our approximate Hamiltonian as polynomials in these unperturbed creation and annihilation operator fields. This Hamiltonian is expressed as the sum of an unperturbed Hamiltonian which is quadratic in the free field creation and annihilation operators and a perturbation term which contains cubic and higher degree terms in the same. By going over to the interaction picture, or equivalently the Dyson series, we obtain the approximate time evolution operator in the interaction picture in terms of the free field creation and annihilation operators using which the scattering probability amplitudes between initial and final states are evaluated.

4.7 LDP problems in quantum mechanical transitions

The perturbing Hamiltonian is small and this smallness is characterized by a perturbation parameter. Using standard quantum mechanical time dependent perturbation theory or equivalently, the Dyson series expansion, the problem is to calculate the probability of the system making transition from one stationary state of the unperturbed system to another in terms of the perturbation parameter. In the limit as the perturbation parameter converges to zero, this transition probability will also converge to zero and the problem is to determine the rate at which its convergence takes place.

Let H_0 be the unperturbed Hamiltonian which is a quadratic function of the bosonic and Fermionic annihilation and creation operator fields and $\epsilon V(t)$ the perturbing Hamiltonian which contains cubic and higher order terms in the creation and annihilation operators. The total Hamiltonian is

$$H(t) = H_0 + \epsilon V(t)$$

The Hamiltonian in the interaction picture is

$$\epsilon \tilde{V}(t), \tilde{V}(t) = exp(itH_0)V(t).exp(-itH_0)$$

This interaction Hamiltonian can be obtained from $\epsilon V(t)$ by replacing each annihilation operator $a(k)$ with $a(k)exp(-i\omega(k)t)$ and the corresponding creation operator $a(k)^*$ with $a(k)^*exp(i\omega(k)t)$ for some appropriate frequency $\omega(k)$. The transition probability between two stationary states $|n>, |m>$ of H_0 with $n \neq m$ under the influence of the interactions and the nonlinear terms is then given by

$$P_\epsilon(T, n, m) = |<m|T\{exp(-i\epsilon \int_0^T \tilde{V}(t)dt)\}|n>|^2$$

and for $n \neq m$, this converges to zero as $\epsilon \to 0$ because two different eigenstates of H_0 are orthogonal. A large deviation result will tell us the rate at which this probability converges to zero. The probability amplitude for the above transition can be expressed as

$$A_\epsilon(T, n, m) =<m|T\{exp(-i\epsilon \int_0^T \tilde{V}(t)dt)\}|n>=$$

$$= \sum_{r \geq 1} (-i\epsilon)^r \int_{0<t_r<...<t_1<T} <m|\tilde{V}(t_1)...\tilde{V}(t_r)|n> dt_1...dt_r$$

The LDP problem is whether this transition probability can be expressed as $exp(-I(T, m, n)/\epsilon)$ as $\epsilon \to 0$, ie, whether

$$-\epsilon.log(|A_\epsilon(T, n, m)|^2)$$

converges to a finite positive limit as $\epsilon \to 0$ or more generally, whether there exists a function $g(\epsilon)$ that is positive, decreases to zero and is such that $-g(\epsilon).log(|A_\epsilon(T, n, m)|$ converges to a finite positive limit as $\epsilon \to 0$.

4.8 Large deviation principle for quantum Gaussian states in infinite dimensions using quantum moment generating functions, quantum Gaussian states obtained by perturbing a harmonic oscillator Hamiltonian by small anharmonic terms

Suppose $a(f)$ is an annihilation field on $L^2(\mathbb{R})$. Thus,

$$[a(f), a(g)^*] =< f, g >, f, g \in L^2(\mathbb{R}_+)$$

Let ρ be a Gaussian state. Then we have by definition of Gaussian state that

$$Tr(\rho.W(f)) = exp(L(f) + Q(f))$$

where L is a linear functional and Q is a quadratic functional on $L^2(\mathbb{R}_+)$. Here, $W(f)$ is the Weyl operator defined by

$$W(f) = exp(a(f) - a(f)^*)$$

It follows therefore that

$$Tr(\rho.W(\epsilon.f)) = exp(\epsilon.L(f) + \epsilon^2 Q(f))$$

from which we obtain a large deviation result of the following form for Gaussian states:

$$lim_{\epsilon\to\infty}\epsilon^{-2}(log(Tr(\rho.W(\epsilon f))) - \epsilon.L(f)) = Q(f)$$

The Legendre transform of this limiting shifted logarithmic moment generating function is again a quadratic form:

$$I(g) =< f, g > -Q(f)$$

where f is the extremal point of $\psi(h) =< f, h > -Q(h)$, ie,

$$f = Q'(g)$$

For Gaussian states, we thus have a definite large deviation principle. For non-Gaussian states do we have some approximate version of such a result ? Let a, a^* be a single annihilation-creation pair so that $[a, a^*] = 1$. Consider the Hamiltonian

$$H = H_0 + \epsilon V, H_0 = a^* a, V = f(a, a^*)$$

The Gibbs state corresponding to this Hamiltonian is a weakly non-Gaussian state:

$$\rho = \rho_\epsilon(\beta) = Z(\beta)^{-1}.exp(-\beta H), Z(\beta) = Tr(exp(-\beta H))$$

The quantum characteristic function of this non-Gaussian state is given by

$$\hat{\rho}(z) = Tr(\rho.exp(\bar{z}a - za^*))$$

and the quantum moment generating function is

$$\hat{\rho}(iz) = Tr(\rho.exp(-i\bar{z}a + iza^*))$$

We can evaluate this upto $O(\epsilon)$ terms. In the limit as $\epsilon \to 0$, this state converge to the Gibbs state associated with the unperturbed Hamiltonian. At what rat the probability distribution of an observable in this state converge to that the same observable in the unperturbed state ?

4.9 Large deviation problems in queueing theory

[1] $M/M/1$ queue. Poisson arrivals with rate λ and Poisson service with rat μ, ie, the inte-arrival times are iid exponential r.v's with mean $1/\lambda$ and th service times are iid exponential r.v's with mean $1/\mu$. Let $P(t,n), n \geq 0$ be th probability that at time t, the queue length is n. Then, we have

$$P(t+dt,n) = P(t,n-1)\lambda dt + P(t,n+1)\mu dt + P(t,n)(1-(\lambda+\mu)dt), n \geq 1$$

so that

$$dP(t,n)/dt = \lambda P(t,n-1) + \mu P(t,n+1) - (\lambda+\mu)P(t,n), n \geq 1$$

For $n = 0$, we have

$$P(t+dt,0) = P(t,1)\mu dt + P(t,0)(1-\lambda dt)$$

so that

$$dP(t,n)/dt = \mu P(t,1) - \lambda P(t,0)$$

Let $P(n), n \geq 0$ denote the steady state probability distribution. Then,

$$\lambda P(n-1) + \mu P(n+1) - (\lambda+\mu)P(n) = 0, n \geq 1,$$

$$\mu P(1) - \lambda P(0) = 0$$

These equations can be arranged as

$$(\lambda/\mu)(P(n) - P(n-1)) = (P(n+1) - P(n)), n \geq 1,$$

$$(\lambda/\mu)P(0) = P(1)$$

The solution is

$$P(n) - P(n-1) = C.(\lambda/\mu)^n, n \geq 1$$

$$P(n) = P(0) + \sum_{m=0}^{n}(P(m) - P(m-1)) = C_1 + C_2(\lambda/\mu)^n, n \geq 0$$

Obviously $C_1 = 0$ since $P(n)$ is summable. This gives

$$P(n) = C_1(\lambda/\mu)^n, n \geq 0$$

and the normalization

$$\sum_{n \geq 0} P(n) = 1$$

then gives

$$C_1 = (1 - \lambda/\mu)$$

Thus,

$$P(n) = (1 - \lambda/\mu)(\lambda/\mu)^n, n \geq 0$$

and we must obviously assume that $\lambda \leq \mu$. From an intuitive standpoint, this must indeed be so for if the arrival rate is greater than the service rate, the queue length will grow indefinitely with time and no stationary distribution can exist.

[2] Little's theorem: Let T_n be the arrival time of the nth packet, W_n the waiting time of the nth packet and S_n the service time of the nth packet. $\tau_{n+1} = T_{n+1} - T_n, n \geq 0$ are assumed to be iid with distribution $F(x)$ while $\{S_n, n \geq 1\}$ are iid and independent of $\{\tau_{n+1} : n \geq 0\}$ with distribution $G(x)$. The waiting times $W_n, n \geq 0$ are easily seen to satisfy the recursion

$$W_{n+1} = max(T_n + W_n + S_n - T_{n+1}, 0) = max(W_n + X_{n+1}, 0)$$

where

$$X_{n+1} = S_n - \tau_{n+1}$$

Note that $X_{n+1}, n \geq 0$ are iid with distribution

$$H(x) = \int_0^\infty G(x+y)dF(y), x \in \mathbb{R}$$

Then it follows that $W_n, n \geq 0$ is a Markov chain with state space $[0, \infty)$ and with stationary transition probabilities given by

$$P(W_{n+1} \leq v | W_n = w) = P(-w < X_{n+1} \leq v - w) + P(X_{n+1} \leq -w)$$

$$= H(v - w), v, w \in [0, \infty)$$

Then, if Φ_n denotes the distribution of W_n, then we get

$$\Phi_{n+1}(v) = \int H(v - w)d\Phi_n(w)$$

and hence if Φ is the stationary/equilibrium distribution of W_n, then

$$\Phi(v) = \int_0^\infty H(v-w)d\Phi(w), v \geq 0$$

This is the Wiener-Hopf integral equation and can be solved by standard methods.

[3] The embedded Markov chain method for analyzing queues: Let $n(i)$ denote the number of units in the queue just after the ith departure and let $a(i+1)$ denote the number of arrivals during the service time of the $(i+1)^{th}$ unit. If S_{i+1} denotes the service time of the $(i+1)^{th}$ unit and if $G(t)$ is the probability distribution of the service times, then we have the obvious recursion

$$n(i+1) = n(i) - 1 + a(i+1)$$

if $n(i) > 0$ and if $n(i) = 0$, then

$$n(i+1) = a(i+1)$$

This gives us

$$\mathbb{E}[z^{n(i+1)}] = \mathbb{E}[z^{a(i+1)}]P(n(i)=0) + \mathbb{E}[z^{a(i+1)}].\mathbb{E}[z^{n(i)-1}|n(i)>0]$$

Now if the arrival times are exponential with mean $1/\mu$, then

$$\mathbb{E}[z^{a(i+1)}] = \sum_{n\geq 0} z^n P(N(S_{i+1}) = n)$$

$$= \sum_{n\geq 0} z^n \int P(S_{i+1} \in dt)P(N(t) = n)$$

$$= \sum_{n\geq 0} z^n \int exp(-\mu t)((\mu t)^n/n!)dG(t)$$

$$= \int exp(\mu t(z-1))dG(t) = M_S(\mu(z-1))$$

where

$$M_S(s) = xp(s.S_{i+1}) = \int_0^\infty exp(st)dG(t)$$

is the moment generating function of each service time.

[4] Large deviation problems: If the mean arrival and mean service times in a queue are themselves random variables, or more specifically, suppose we consider an $M/M/1$ queue with infinitesimal transition probabilities

$$P(X(t+dt) = i+1|X(t) = i) = \lambda(i)dt, i \geq 0,$$

$$P(X(t+dt) = i-1|X(t) = i) = \mu(i)dt, i \geq 1,$$

$$P(X(t+dt) = i|X(t) = i) = 1-(\lambda(i)+\mu(i))dt, i \geq 1,$$

$$P(X(t+dt) = 0|X(t) = 0) = 1-\lambda(0)dt$$

where $\{\lambda(i), i \geq 0\}$ are iid, $\{\mu(i), i \geq 1\}$ are also iid and further, these two sets of r.v's are mutually independent of each other, then we have a "queue in a random environment" and we can formulate large deviation problems for such queues, like for example ask the following question: Does $X(t)/t$ converge to a definite random variable as $t \to \infty$ and if so, then at what rate ? How does the empirical density $T^{-1}\int_0^T \delta_{X(t)}dt$ of $X(.)$ behave as $T \to \infty$? Does it have a rate function ? For this, we must first assume that the initial condition $X(0)$ is chosen so that $X(.)$ is a stationary process, ie, if $P(t,n) = P(X(t) = n|\lambda, \mu), n \geq 0$, then we have

$$\partial_t P(t,n) = P(t,n+1)\mu(n+1)+P(t,n-1)\lambda(n-1)-P(t,n)(\lambda(n)+\mu(n)), n > 1,$$

$$\partial_t P(t,0) = P(t,1)\mu(1) - P(t,0)\lambda(0)$$

and then we require that $P(0,n)$ satisfy

$$0 = P(0,n+1)\mu(n+1) + P(0,n-1)\lambda(n-1) - P(0,n)(\lambda(n)+\mu(n)), n \geq 1,$$

$$0 = P(0,0)\lambda(0) + P(0,1)\mu(1) - P(0,0)\lambda(0)$$

so that we are guaranteed that conditioned on the spatial process $\lambda(.), \mu(.)$, $X(.)$ is a stationary process. It is easily seen then that when the conditioning is removed, the process remains stationary. Note that the above conditional equilibrium equations can be expressed as (with $P(0,n) = P(n)$),

$$\lambda(n-1)P(n-1) - \mu(n)P(n) = \lambda(n)P(n) - \mu(n+1)P(n+1), n \geq 1,$$

$$\lambda(0)P(0) = \mu(1)P(1)$$

so that

$$\lambda(n)P(n) - \mu(n+1)P(n+1) = c, n > 0$$

where c is a constant random variable. Taking $c = 0$ gives

$$P(n+1)/P(n) = \lambda(n)/\mu(n+1)$$

and hence

$$P(n) = P(0)\Pi_{k=0}^{n-1}[\lambda(k)/\mu(k+1)], n \geq 1$$

$P(0)$ is then determined from the normalization condition

$$\sum_{n \geq 0} P(n) = 1$$

and thus we have an expression for $P(n)$ as a nonlinear function of the r.v's $\lambda(.), \mu(.)$. For the equilibrium distribution, to get

$$Q(t,n) = P(X(t) = n) = \mathbb{E}P(t,n)$$

we must take the expectation of $P(t,n)$ w.r.t the r.v's $\lambda(.), \mu(.)$. It is clear that LDP properties of the random probabilities $P(t,n)$ will be decided by the LDP properties of the sum $\sum_{k=0}^{n-1} log(\lambda(k)/\mu(k+1))$. Since this is a sum of iid r.v's, we can apply Cramer's theorem to this sum.

[5] Consider the above example of a queue in a random environment but now assume that the queue number can also assume all negative integer values. In other words, we have a birth-death process with $\mathbb{Z}$ as the state space and the process being continuous in time. The infinitesimal transition probabilities are

$$P(X(t+dt) = n+1|X(t) = n) = \lambda(n)dt, P(X(t+dt) = n-1|X(t) = n) = \mu(n)dt,$$

$$P(X(t+dt) = n|X(t) = n) = 1 - (\lambda(n) + \mu(n))dt$$

for all $n \in \mathbb{Z}$. This Markov chain chain can be regarded as a continuous time version of a discrete random walk. We now assume that $\{(\lambda(n), \mu(n)) : n \in \mathbb{Z}\}$ is a stationary random process on discrete space $\mathbb{Z}$. The question is what is the large deviation rate function for the averaged chain, ie, for the empirical distributions

$$T^{-1}\int_0^T \delta_{X(t)}dt$$

For this we have to first ascertain that the limiting logarithmic moment generating function

$$\Lambda(f) = lim_{T\to\infty}T^{-1}.log\mathbb{E}[exp(\int_0^T f(X(t))dt)]$$

exists. Note that the averaged process $X(t)$ can be stationary but is not Markov.

4.10 Large deviation problems associated with quantum filtering theory

Consider a cavity electromagnetic field having N modes, ie, N frequencies of oscillation. This electromagnetic field can be expressed as a Hermitian superposition of N independent creation and annihilation operator pairs $a(k)^*, a(k), k = 1, 2, ..., N$ satisfying the CCR $[a(k), a(m)^*] = \delta(k,m)$. We can write down the magnetic vector potential of this quantum field as

$$\mathbf{A}_S(t,\mathbf{r}) = \sum_{k=1}^{N}[a(k)\mathbf{f}_k(t,\mathbf{r}) + a(k)^*\mathbf{f}_k(t,\mathbf{r})^*]$$

$$\Phi_S(t,\mathbf{r}) = \sum_{k=1}^{N}[a(k)g_k(t,\mathbf{r}) + a(k)^*g_k(t,\mathbf{r})^*]$$

here by the Lorentz gauge condition

$$div\mathbf{f}_k(t,\mathbf{r}) = -\partial_t g_k(t,\mathbf{r}), k = 1,2,...,N$$

he interaction energy between this quantum electromagnetic field and an atom ith an electron located at $\mathbf{r}_0$ within the cavity is given by

$$I_I(t) = (e/2m)(\mathbf{p},\mathbf{A}(t,\mathbf{r}_0+\xi))+(\mathbf{A}_S(t,\mathbf{r}_0+\xi),\mathbf{p})-e\Phi_S(t,\mathbf{r}_0+\xi) \\ -ge(\mathbf{B}_S(t,\mathbf{r}_0+\xi),\sigma)/2m$$

here ξ is the position observable of the electron, σ is the spin observable vector f the electron and $\mathbf{p} = -i\nabla_\xi$ is the momentum observable of the electron and

$$\mathbf{B}_S(t,\mathbf{r}) = curl\mathbf{A}_S(t,\mathbf{r}) = \sum_k [a(k)curl\mathbf{f}_k(t,\mathbf{r}) + a(k)^* curl\mathbf{f}_k(t,\mathbf{r})^*]$$

s the magnetic field of the cavity system electromagnetic field. The atomic Iamiltonian is

$$H_A = \mathbf{p}^2/2m - Ze^2/|\xi|$$

he total system Hilbert space consists of the tensor product of (a) the electron osition Hilbert space $L^2(\mathbb{R}^3)$, (b) the electron spin Hilbert space $\mathbb{C}^2$ and the avity field Hilbert space $L^2(\mathbb{R})^{\otimes N} = L^2(\mathbb{R}^N)$, where by $=$, we really mean Iilbert space isomorphism. The corresponding system Hamiltonian is given by

$$H_S(t) = H_A + H_I(t) + H_F$$

here H_F is the cavity field Hamiltonian given by

$$H_F = \sum_{k=1}^N \omega(k)a(k)^* a(k)$$

Iote that we can equivalently express the system Hamiltonian as

$$I_S(t) = (1/2m)(\mathbf{p}+e\mathbf{A}(t,\mathbf{r}_0+\xi))^2 - e\Phi(t,\mathbf{r}_0+\xi) \\ -(ge/2m)(\sigma,\mathbf{B}_S(t,\mathbf{r}_0+\xi)) + \sum_k \omega(k)a(k)^* a(k)$$

he bath Hamiltonian that surrounds the cavity is described by the bath magetic vector potential

$$\mathbf{A}_B(t,\mathbf{r}) - \mathbf{A}'(t)\mathbf{F}(\mathbf{r}) + \mathbf{A}'(t)^*\mathbf{F}(t,\mathbf{r})^* + \Lambda'(t)\mathbf{G}(\mathbf{r})$$

here $\mathbf{A}(t), \mathbf{A}(t)^*, \Lambda(t), t \geq 0$ are the annihilation, creation and conservation rocesses of the Hudson-Parthasarathy quantum stochastic calculus. $\mathbf{A}(t)$ The orentz gauge condition

$$div\mathbf{A}_B(t,\mathbf{r}) = -\partial_t \Phi_B(t,\mathbf{r})$$

ives the bath electric potential as

$$\Phi_B(t,\mathbf{r}) = -\mathbf{A}(t)div\mathbf{F}(\mathbf{r}) - \mathbf{A}(t)^*.div\mathbf{F}(\mathbf{r}) - \Lambda(t)div\mathbf{G}(\mathbf{r})$$

The total electric field of cavity system and bath is given by

$$\mathbf{E}(t,\mathbf{r}) = \mathbf{E}_S(t,\mathbf{r}) + \mathbf{E}_B(t,\mathbf{r})$$

and the total magnetic field of cavity system and bath is

$$\mathbf{B}(t,\mathbf{r}) = \mathbf{B}_S(t,\mathbf{r}) + \mathbf{B}_B(t,\mathbf{r})$$

where

$$\mathbf{E}_S = -\nabla\Phi_S - \partial_t\mathbf{A}_S$$

is the cavity system electric field,

$$\mathbf{B}_S = curl\mathbf{A}_S,$$

the cavity system magnetic field, has already been defined above and

$$\mathbf{B}_B = curl\mathbf{A}_B = A'(t)curl\mathbf{F}(\mathbf{r}) + A'(t)^* curl\mathbf{F}(\mathbf{r})^* + \Lambda'(t)curl\mathbf{G}(\mathbf{r})$$

is the bath magnetic field and

$$\mathbf{E}_B = -\nabla\Phi_B - \partial_t\mathbf{A}_B$$

is the bath electric field.

Another way to do this calculation is to assume the Coulomb gauge so tha the system and bath electric potentials are zero and the system and bath mag netic vector potentials have vanishing divergences. Then it is more convenier to assume that the bath magnetic vector potential has the form

$$\mathbf{A}_B(t,\mathbf{r}) = \mathbf{A}(t)\mathbf{F}(\mathbf{r}) + \mathbf{A}(t)^*\mathbf{F}(t,\mathbf{r})^* + \Lambda(t)\mathbf{G}(\mathbf{r})$$

and then the bath electric field has the form

$$\mathbf{E}_B = -\partial_t\mathbf{A}_B = -\mathbf{A}'(t)\mathbf{F}(\mathbf{r}) - \mathbf{A}'(t)^*\mathbf{F}(t,\mathbf{r})^* - \Lambda'(t)\mathbf{G}(\mathbf{r})$$

while the system electric field is

$$\mathbf{E}_S = -\partial_t\mathbf{A}_S = -\sum_k [a(k)\partial_t\mathbf{f}_k(t,\mathbf{r}) + a(k)^*\partial_t\mathbf{f}_k(t,\mathbf{r})^*]$$

and then the interaction Hamiltonian between the cavity system and bath elec tric fields is given by

$$H_{BS}(t) = \int(\mathbf{E}_S(t,\mathbf{r}),\mathbf{E}_B(t,\mathbf{r}))d^3r$$

$$= L_1(t)dA(t) + L_1(t)^*dA(t)^* + S(t)d\Lambda(t)$$

where

$$L_1(t) = -\int(\mathbf{E}_S(t,\mathbf{r}), F(\mathbf{r})d^3r,$$

$$S(t) = -\int (\mathbf{E}_S(t, \mathbf{r}), \mathbf{G}(\mathbf{r}))d^3r$$

Note that $L(t), S(t)$ are functions of the cavity field and atomic observables: $(k), a(k)^*, \xi, \mathbf{p}$. Note that $H_{BS}(t)$ is a bilinear combination of $(A'(t), A'(t)^*, \Lambda'(t))$ nd $(a(k), a(k)^*)_{k=1}^N$. Note that in this Coulomb gauge, we must have

$$div\mathbf{f}_k(t, \mathbf{r}) = 0 = div\mathbf{F}(\mathbf{r})$$

The total Hamiltonian of the cavity with atom system interacting with the bath s

$$H(t) = H_S + H_{BS} = H_F + H_A + H_I + H_{BS}$$

nd we can cast the Schrodinger equation for evolution under this Hamiltonian f system $\otimes$ bath as a quantum stochastic differential equation with quantum Ito orrection terms $P(t)$ added to it in order to guarantee unitary evolution. Let $_t(X)$ denote the evolution of a Heisenberg system observable under this dynamcs and let $\pi_t(X)$ denote its conditional expectation given the non-demolition neasurements $\eta(t)$ upto time t. Note that the measurement model has the form

$$Y_o(t) = U(t)^* Y_i(t) U(t), Y_i(t) = cA(t) + \bar{c}A(t)^*$$

where

$$dU(t) = -i(H(t)dt + P(t)dt)U(t)$$

$$H(t)dt = (H_F + H_A + H_I(t))dt = (H_F + H_A)dt + L_1(t)dA(t) + L_2(t)dA(t)^* + S(t)d\Lambda(t)$$

nd so

$$dY_o(t) = j_t(M_1(t))dt + j_t(M_2(t))dA(t) + j_t(M_2(t)^*)dA(t)^* + j_t(M_3(t))d\Lambda(t)$$

where M_1, M_2, M_3 are system observables. This measurement model should e compared to the situation in classical nonlinear filtering theory where the neasurement has the form

$$dz(t) = h(x(t))dt + dv(t)$$

When the process and measurement noise processes are small, we incorporate mall perturbation parameters ϵ into them. This amounts to replacing L_1 by $\sqrt{\epsilon}L_1$ and S by $\epsilon S(t)$. Let us consider a classical process

$$Z(t, \epsilon) = \sqrt{\epsilon}B(t) + \epsilon N(\epsilon, t)$$

where $B(t)$ is a standard Brownian motion process and $N(\epsilon, t)$ is a Poisson rocess with rate $\lambda(\epsilon)$. The moment generating functional of this process is

$$exp(\Lambda(f, \epsilon)) = \mathbb{E}(exp(\int_0^T f(t)dZ(t, \epsilon)))$$

$$= exp((\epsilon/2)\int_0^T f^2(t)dt).exp(\lambda(\epsilon)\int_0^T (exp(\epsilon f(t)) - 1)dt)$$

nd hence the Gartner-Ellis limiting logarithmic moment generating functional s given by

$$\Lambda(f) = lim_{\epsilon \to 0}\epsilon.log[\Lambda(\epsilon^{-1}f, \epsilon)]$$

$$= (1/2)\int_0^T f^2(t)dt + \lambda_0\int_0^T (exp(f(t)) - 1)dt$$

where we assume that the limit

$$\lambda_0 = lim_{\epsilon\to 0}\epsilon.\lambda(\epsilon)$$

exists, which means that as $\epsilon \to 0$, the Poisson arrival rate becomes infinite. In the quantum case, consider first the rate function for the quantum Brownian motion in a coherent state. The moment generating functional is

$$M(f) =< \phi(u)|exp(\int_0^T f(t)(cdA(t) + \bar{c}dA(t)^*))|\phi(u) >=$$

$$M(f,u) = + < \phi(u)|exp(ca(f) + \bar{c}a(f)^*)|\phi(u) >$$

where f is assumed to be a real valued function. This evaluates to

$$exp(|c|^2 < f,f > /2).exp(2Re(c < f,u >))$$

The rate function $I(x)$ is then the Legendre transform of the logarithmic moment generating function $\Lambda(f) = |c|^2 < f,f > /2 + 2.Re(c < f,u >)$:

$$I(x) = sup_f(< f,x' > -\Lambda(f) >)$$

$$= sup_f(< f,x' > -|c|^2 < f,f > /2-2.Re(c < f,u >))$$

where $x' = dx/dt$. Setting the variational derivative w.r.t f to zero gives the following equation for the optimal f:

$$x' - |c|^2 f - 2.Re(cu) = 0$$

Thus, for this optimal f,

$$< f,x' >= |c|^2 < f,f > +2Re(c < f,u >)$$

and hence

$$I(x) = |c|^2 < f,f > /2 =\| x' - 2Re(cu) \|^2 /|c|^2$$

This is the rate function for the quantum Brownian motion $cA(t) + \bar{c}A(t)^*$ in the coherent state $|\phi(u) >$. Now let us look at the same for a quantum Poisson process $\Lambda_t(H) = \lambda(H_t)$. We have

$$\int_0^T f(t)d\Lambda_t(H) = \lambda(Hf_T)$$

where $f_T(t) = f(t)\chi_{[0,T]}(t)$. Then

$$< \phi(u)|\int_0^T f(t)d\Lambda_t(H)|\phi(u) >=< \phi(u)|exp(\lambda(Hf_T)|\phi(u) >=$$

$$= exp(< u|(exp(Hf_T) - 1)|u >)$$

Note that here the Boson Fock space is $\Gamma_s(\mathcal{H}_0 \otimes L^2(\mathbb{R}_+))$, H is a self-adjoint operator in $\mathcal{H}_0$ and the standard spectral family of indicators $\chi_{[0,t} >, t \geq 0$ acts in $L^2(\mathbb{R}_+)$. We write the spectral representation of H as

$$H = \int_{\mathbb{R}} xdE(x)$$

Then, the logarithmic moment generating function of the process $\Lambda_t(H), t \in [0, T]$ in the coherent state $|\phi(u) >$ is given by

$$< u|(exp(Hf_T)-1)|u >= \int < u|(exp(xf_T)-1)dE(x)|u >$$

$$= \int_{\mathbb{R}\times[0,T]} (exp(xf(t))-1)d_x < u(t)|E(x)|u(t) > dt$$

where

$$u \in \mathcal{H}_0 \otimes L^2(\mathbb{R}_+)$$

or equivalently,

$$u(t) \in \mathcal{H}_0, t \in \mathbb{R}_+$$

It is clear from this formula that $\Lambda_t(H)$ is a compound Poisson process in the coherent state $|\phi(u) >$. In fact writing

$$\mu(t, x) = d < u(t)|E(x)|u(t) > /dx$$

we see that

$$log(< \phi(u)| \int_0^T f(t)d\Lambda_t(H)|\phi(u) >) = \int_{\mathbb{R}\times[0,T]} (exp(xf(t)) - 1)\mu(t, x)dxdt$$

In case $H = c|v >< v|$ is a unit rank operator with $|v >$ a unit vector, this reduces to the logarithmic generating function of a Poisson process:

$$log(< \phi(u)| \int_0^T f(t)d\Lambda_t(H)|\phi(u) >) = \int_0^T (exp(cf(t)) - 1)\mu(t)dt$$

where

$$\mu(t) = | < u(t)|v > |^2$$

This means that the usual large deviation principle can be applied to this quantum Poisson process as well as to the previous compound Poisson process.

Chapter 5

LDP in Classical Stochastic Process Theory and Quantum Mechanical Transitions

5.1 Large deviations problems to the propagation of noise at the sigmoidal computation nodes through the neural network

Consider a p-layered network with each layer having L nodes. The input $x_0(t)$ is applied to the first layer and let $x_k(t)$ denote the signal vector at the kth layer. Thus assuming feed-forward dynamics, we have the recursions

$$x_{k+1}(t) = \sigma(W_{k+1}(t)x_k(t)) + v_{k+1}(t), k = 0, 1, ..., p-1$$

The output is taken at the pth layer, ie,

$$y(t) = x_p(t) = \sigma(W_p(t)x_{p-1}(t)) + v_p(t)$$

Here, the $v_k(t)'s$ are small amplitude noise processes. Now we design the weight adapation using the EKF so that the dynamics of its EKF estimate is derived from noisy measurements of the output process for a given input process. In principle, we can assume that we have solved for the weight estimate process over a time interval $[0, T]$. Denote this weight estimate process by $\hat{W}(t)$. We can even do this for a recurrent neural network when the nodal state process follows a differential equation

$$dX(t)/dt = F(X(t), W(t), U(t))$$

where $U(t)$ is the applied input. The weight process then satisfies the sde

$$dW(t) = \sigma.dB(t)$$

where $B(.)$ is vector valued Brownian motion. The extended state $[X(t), W(t)]$ is estimated using the EKF based on noisy output measurements $Z(t)$ and we denote it by $\hat{W}(t)$. This EKF is of the form

$$d[\hat{X}(t)^T, \hat{W}(t)^T]^T = [F(\hat{X}(t), \hat{W}(t), U(t))^T, 0]^T dt + K(t)(dZ(t) - H\hat{X}(t)dt)$$

where

$$dZ(t) = Y(t)dt + dV(t)$$

with $Y(t)$ being the desired output vector of the neural network and $HX(t)$ being the true ouptut vector of the neural network.. The LDP problem is to calculate the rate function of the weight estimate process $\hat{W}(t), t \in [0, T]$ when the noise processes $B(.)$ and $V(.)$ are weak amplitude and hence to compute the approximate probability that for a given input the nn output given these weights will generate an output that deviates from the desired one by an amount greater than a given threshold.

5.2 Law of the iterated logarithm for sums of iid random variables

Let $X_n, n \geq 1$ be iid finite variance r.v's with zero mean and unit variance. Consider for $a > 1$

$$S_n = \sum_{k=1}^{n} X_k, M_n = sup(S_k : 1 \leq k \leq n)$$

We have

$$\{M_n/\sqrt{n} \geq x\} = \{S_n/\sqrt{n} \geq x-a\} \cup \bigcup_{k=1}^{n} \{S_n/\sqrt{n} \leq x-a, M_{k-1}/\sqrt{n} < x \leq M_k/\sqrt{n}\}$$

Now

$$\{S_n/\sqrt{n} \leq x-a, M_{k-1}/\sqrt{n} < x \leq M_k/\sqrt{n}\} \subset \{S_n/\sqrt{n} \leq x-a, S_k/\sqrt{n} \geq x, M_{k-1}/\sqrt{n} < x \leq M_k/\sqrt{n}\}$$

$$\subset \{(S_n-S_k)/\sqrt{n} \leq -a, M_{k-1}/\sqrt{n} < x \leq M_k/\sqrt{n}\} \subset \{|S_n-S_k|/\sqrt{n} \geq a, M_{k-1}/\sqrt{n} < x \leq M_k/\sqrt{n}\}$$

and hence

$$P(M_n/\sqrt{n} \geq x) \leq P(S_n/\sqrt{n} \geq x-a)+$$

$$P(M_{k-1}/\sqrt{n} < x \leq M_k/\sqrt{n}, |S_n-S_k|/\sqrt{n} \geq a)$$

Now, by Chebyshev's inequality,

$$P(|S_n - S_k|/\sqrt{n} \geq a) \leq (n-k)/a^2 n \leq 1/a^2$$

and further the events $\{|S_n - S_k|/\sqrt{n} \geq a\}$ and $\{M_{k-1}/\sqrt{n} < x \leq M_k/\sqrt{n}\}$ are independent because $S_n - S_k$ is independent of $\{M_{k-1}, M_k\}$. This gives

$$P(M_n/\sqrt{n} \geq x) \leq P(S_n/\sqrt{n} \geq x-a)+(1/a^2)\sum_{k=1}^{n} P(M_{k-1}/\sqrt{n} < x \leq M_k/\sqrt{n})$$

$$= P(S_n/\sqrt{n} \geq x - a) + (1/a^2)P(M_n/\sqrt{n} \geq x)$$

This implies that

$$P(M_n/\sqrt{n} \geq x) \leq (1 - 1/a^2)^{-1}P(S_n/\sqrt{n} \geq x - a)$$

This inequality is at the heart of the derivation of the law of the iterated logarithm for sums of iid zero mean finite variance random variables.

5.3 A version of the LDP for iid random variables

Let $0 < a(n) \to \infty$ be such that $a(n)/\sqrt{n} \to 0$ and let $S_n = X_1 + ... + X_n$ be the sum of n iid r.v's each having mean zero and unit variance. Consider the r.v's

$$Z_n = S_n/a(n)\sqrt{n}, n \geq 1$$

We compute the Gartner-Ellis limiting logarithmic moment generating function

$$a(n)^{-2}.log\mathbb{E}exp(s.a(n)^2 Z_n)$$

$$= na(n)^{-2}.log(M_X(sa(n)/\sqrt{n}) = na(n)^{-2}\Lambda_X(sa(n)/\sqrt{n})$$

$$= na(n)^{-2}(s^2a(n)^2/2n + o(a(n)^2/n)$$

As $n \to \infty$, this converges to $s^2/2$ and its Legendre transform is $x^2/2$. Thus, we get from the LDP,

$$P(Z_n > x) \approx exp(-a(n)^2x^2/2)$$

or more precisely,

$$lim_{n\to\infty}a(n)^{-2}.log(P(Z_n > x)) + x^2/2 = \xi(n) \to 0$$

so that

$$P(Z_n > x) = exp(-a(n)^2(x^2/2 + \xi(n)))$$

where

$$\xi(n) \to 0$$

5.4 The law of the iterated logarithm for sums of iid random variables having finite variance

Let $X_n, n \geq 1$ be iid r.v's with $\mathbb{E}X_1 = 0, \mathbb{E}(X_1^2) = 1$ and define $S_n = X_1 + ... + X_n$. We wish to show that

$$P(limsup_n S_n/h(n) = 1), h(n) = \sqrt{2n.log(log(n))}$$

Note that this result is a stronger version of the same law for Brownian motion because the increments of Brownian motion are normally distributed whereas here we do not assume any specific law for the increments of the process S_n except that the increments are independent and have a variance proportional to the number of $X_i's$ in these increments.

5.5 An open problem relating applications of LDP to Martingales

Let $X(n)$ be a Martingale in discrete time. Assuming that $sup_n \mathbb{E}(|X(n)|) < \infty$ show that $lim_{n\to\infty} X(n) = X(\infty)$ exists and that $X(\infty)$ is integrable and finally under the additional assumption that $X(n), n \geq 0$ is uniformly integrable, that

$$X(n) = \mathbb{E}(X(\infty)|\mathcal{F}_n), n \geq 0$$

where

$$\mathcal{F}_n = \sigma(X(k) : k \leq n)$$

Suppose $X(n)$ is a submartingale w.r.t to the filtration $\mathcal{F}_n$. Define

$$\Delta A(n) = \mathbb{E}(X(n) - X(n-1)|\mathcal{F}_{n-1}) = \mathbb{E}(X(n)|\mathcal{F}_{n-1}) - X(n-1)$$

Clearly $\Delta A(n) \geq 0$. Define

$$A(n) = \sum_{k \leq n} \Delta A(k)$$

Then $A(n)$ is increasing and further if

$$\Delta M(n) = \Delta X(n) - \Delta A(n)$$

where

$$\Delta X(n) = X(n) - X(n-1)$$

then

$$\mathbb{E}(\Delta M(n)|\mathcal{F}_{n-1}) = 0$$

and therefore

$$M(n) = \sum_{k \leq n} \Delta M(k)$$

is a Martingale. Then, we have the Doob-decomposition

$$X(n) = M(n) + A(n)$$

with $M(.)$ a Martingale and $A(.)$ an increasing predictable process w.r.t the filtration $\mathcal{F}_n$. By predictable, we mean that $A(n)$ is $\mathcal{F}_{n-1}$-measurable. Now suppose $W(k)$ is an iid process with zero mean. Consider the Martingale

$$M(n) = \sum_{k \leq n} W(k)$$

We have a large deviation principle for the family

$$M(n)/n, n \to \infty$$

If $W(k)$ is not iid but the condition $\mathbb{E}(W(n)|\mathcal{F}_{n-1}) = 0$, then $M(n) - \sum_{k \leq n} W(k)$ is a Martingale. Do we have any general LDP for the process $M(n)/n$ in this case ? For example, let $X(n)$ be a Markov process with stationary transition probabilities. Write

$$P(x, dy) = P(X(n+1) \in dy|X(n) = x)$$

Then consider the process

$$W(n) = f(X(n)) - f(X(n-1)) - \int P(X(n-1), dy)f(y)$$

$$= f(X(n)) - \mathbb{E}(f(X(n))|X(n-1))$$

Then clearly,

$$\mathbb{E}(W(n)|\mathcal{F}_{n-1}) = 0$$

and hence

$$M(n) = \sum_{k=0}^{n} W(k)$$

is a Martingale. Can one construct an LDP for the process $M(n)/n, n \geq 1$ in terms of the function f and the transition probability $P(x, B)$ of the Markov process $X(.)$? Further, if one drives a discrete time stochastic difference system with the Martingale difference $W(.)$, then does such a process exhibit an LDP? The process Z satisfies the stochastic difference equation

$$Z(n) = \psi(Z(n-1)) + \sqrt{\epsilon}\phi(Z(n-1))W(n)$$

where ϵ is a small parameter that converges to zero.

5.6 Properties of ML estimators based on iid measurements

Let $X_k, k = 1, 2, ...$ be iid with pdf $p(X|\theta)$. Estimate θ based on the observations $X_k, k = 1, 2, ..., N$ using the maximum likelihood method:

$$\hat{\theta}_N = argmax_\theta l_N(\theta)$$

$$l_N(\theta) = N^{-1}.\sum_{k=1}^{N} log(p(X_k|\theta))$$

Using the law of large numbers and the central limit theorem, derive an approximate formula for the probability distribution of $\delta\theta_N = \hat{\theta}_N - \theta$ for large N. For this you can use the approximation

$$l_N(\theta + \delta\theta_N) \approx l_N(\theta) + l'_N(\theta)^T\delta\theta_N + (1/2)\delta\theta_N^T l''_N(\theta)\delta\theta_N$$

and setting the gradient of this w.r.t $\delta\theta$ to zero here gives

$$\delta\theta_N = -l''_N(\theta)^{-1}l'_N(\theta)$$

and then use the law of large numbers to show that for large N,

$$l'_N(\theta) \approx \mathbb{E}\partial(log(p(X|\theta))/\partial\theta) = 0$$

$$Cov(l'_N(\theta)) \approx N^{-2}.Cov(\partial(log(p(X|\theta)/\partial\theta)) = N^{-2}J(\theta)$$

and again by the law of large numbers,

$$l''_N(\theta) \approx \mathbb{E}[\partial^2 log(p(X|\theta))/\partial\theta\theta^T] = -J(\theta)$$

where $J(\theta)$ is the Fisher information matrix and therefore, it follows from the central limit theorem that $\delta\theta_N$ is for large N approximately normal with mean zero and covariance

$$J(\theta)^{-1}.Cov(l'_N(\theta)).J(\theta)^{-1} \approx$$
$$N^{-2}J(\theta)^{-1}$$

[36] Let $X_1, X_2, ..$ be iid with pdf $p_1(X)$ under hypothesis H_1 and iid with pdf $p_0(X)$ under hypothesis H_0. The Neyman-Pearson criterion for testing between these two alternatives is to use the decision regions

$$Z_{1N} = \{(X_1, ..., X_N) : |(1/N)\sum_{k=1}^{N} log(p_1(X_k)/p_0(X_k)) > \eta_N\}$$

$$Z_{0N} = Z_{1N}^c$$

as the decision regions, ie, decide H_1 if $X \in Z_{1N}$ and H_0 otherwise. The threshold η_N is selected so that the false alarm probability

$$\alpha = P(H_1 decided | H_0 true) = \int_{Z_{1N}} (\Pi_{k=1}^{N} p_0(X_k))dX_1...dX_N$$

is a given fixed positive real number lesser than or equal to one. Now applying the LDP, we get for large N,

$$N^{-1} log(\alpha) \approx -inf(I_0(x) : x \in Z_{1N}) = -inf_{x \geq \eta_N} I_0(x)$$

where

$$I_0(x) = sup_{\lambda \in \mathbb{R}}(\lambda.x - \Lambda_0(\lambda))$$

where

$$\Lambda_0(\lambda) = log.\mathbb{E}_0[exp(\lambda.log(p_1(X)/p_0(X)))]$$

$$= log. \int p_0(X)^{1-\lambda} p_1(X)^{\lambda} dX$$

Note that

$$\mu_0 = \mathbb{E}_0(log(p_1(X)/p_0(X)) = \int p_0(X).log(p_1(X)/p_0(X))dX < 0,$$

and

$$\mu_1 = \mathbb{E}_1(log(p_1(X)/p_0(X)) = \int p_1(X).log(p_1(X)/p_0(X))dX > 0$$

Since $\alpha > 0$ is being assumed, it follows that $N^{-1}.log(\alpha) \to 0$ and hence

$$-inf_{x \in Z_{1N}} I_0(x) \to 0$$

Now,

$$I_0(x) = \lambda.x - \Lambda_0(\lambda)$$

where λ satisfies

$$x = \Lambda'(\lambda)$$

Then for this λ, we can write assuming that $\eta_N \geq 0$ (and hence $\eta_N \geq \mu_0$) that

$$-inf_{x \geq \eta_N} sup_{\lambda}(\lambda.x - \Lambda_0(\lambda))$$

$$= -sup_{\lambda \geq 0}(\lambda.\eta_N - \Lambda_0(\lambda)) = -(\lambda_N.\eta_N - \lambda_0(\lambda_N))$$

where λ_N satisfies

$$\eta_N = \Lambda_0'(\lambda_N)$$

This equation is consistent with our hypothesis that $\eta_N \geq 0$ and $\lambda_N \geq 0$ since $\Lambda_0'(0) = \mu_0 < 0$ and $\Lambda_0'(\lambda)$ is an increasing function for all $\lambda \in \mathbb{R}$. Then,

$$log(\alpha)/N \approx -(\lambda_N \Lambda_0'(\lambda_N) - \Lambda_0(\lambda_N))$$

and since this must converge to zero as $N \to 0$ it must follow that $\lambda_N \to ($
Thus, the asymptotic threshold of the test is given by

$$\eta_N \approx \Lambda_0'(\lambda_N) \to \Lambda_0'(0) = \mu_0 = -\int p_0(X)log(p_0(X)/p_1(X))dX = -D(p_0||p_1)$$

[37] Large deviations in quantum cosmology:

[1] Let $g_{\mu\nu}^{(0)}(x)$ be the background classical metric and let $\delta g_{\alpha\beta}(x)$ be th quantum fluctuations in this metric. Express the Hamiltonian of general rela tivity approximately as a quadratic form in the position fields $\delta g_{ab}(x), 1 \leq a \leq$ $b \leq 3$ and in the corresponding momentum fields $\delta\pi_{ab}(x)$ where $1 \leq a \leq b \leq$ and by expanding these position fields within a cube of side length L as a 3-I spatial Fourier series, express the ADM action in terms of a countable set c position variables and their time derivatives with the approximation containin quadratic and even cubic terms. Express using the Legendre transform, the cor responding approximate ADM Hamiltonian as a quadratic form in a countabl set of position and momentum variables with cubic perturbation terms. Assum ing the cubic terms to be small by introducing a small perturbation paramete ϵ into these terms, calculate the approximate transition probability matrix ele ments between two stationary states of the unperturbed Hamiltonian consistin only of quadratic terms as a power series in ϵ and evaluate the LDP rate a which these transition probabilities converge to zero when the cubic perturba tion parameter $\epsilon \to 0$.

[38] **Test on Stochastic processes and queueing theory**

[1] Let $P_n(t) = Pr(X(t) = n), n = 0, 1, 2, ...$ where $X(t)$ is the birth death process with birth rate λ and death rate μ. Write down the Chapman Kolmogorov equations for $P_n(t)$ and hence derive a formula for the generatin function

$$\Phi_N(t, z) = \sum_{n \geq 0} P_n(t) z^n$$

given the initial conditions

$$P_n(0) = \delta(n, N)$$

where N is a fixed positive integer. Hence evaluate $\mathbb{E}(X(t))$ and $Var(X(t))$ Now assume that λ, μ are undergo small random fluctuations, ie,

$$\lambda = \lambda_0 + \delta.\lambda_1, \mu = \mu_0 + \delta.\mu_1$$

where λ_1, μ_1 are independent random variables with a given joint probabilit distribution $F(\lambda_1, \mu_1)$ having a density $f(\lambda_1, \mu_1)$. Calculate the change in th probability $P_n(t)$ before averaging over this random variable pair has done Express the resulting probability as $P_n(t) + \delta.Q_n$. Observe that $Q_n(t), n =$ $1, 2, ...$ is a random function on $\mathbb{Z}_+$. Now apply the LDP to calculate the rat function of $\delta.Q_n(t)$ as $n \to \infty$.

The empirical distribution of the process based on time averages may be computed as

$$T^{-1}\int_0^T (P_n(t)+\delta.Q_n(t))dt$$

and to derive an LDP for this as $T\to\infty$, we must calculate the logarithmic moment generating function

$$T^{-1}log\mathbb{E}exp(\delta\int_0^T Q_n(t)dt)$$

or more generally,

$$T^{-1}.log.\mathbb{E}exp(\delta\sum_n f(n)\int_0^T Q_n(t)dt)$$

and the limit of this as $T\to\infty$. More specifically suppose we are given a usual birth death process $X(t)$. This forms a Markov chain and we know how to compute the rate function for the empirical probability distribution of a Markov chain. If $X(t)$ is a stationary process like say a Markov chain with state space $\mathbb{Z}_+$, then the empirical measure of the multivariate process $(X(t+t_k), k=1,2,...,K), t\geq 0$ is given by

$$\mu_T(B)=T^{-1}\int_0^T \chi_B(X(t+t_k),k=1,2,...,K)dt$$

where

$$B\mathcal{B}(\mathbb{R}^K)$$

and to calculate the rate function of this empirical probability measure, we must evaluate

$$T^{-1}log\mathbb{E}exp(T.\int f(x)\mu_T(dx))=T^{-1}.log\mathbb{E}exp(\int_0^T f(X(t+t_k),k=1,2,...,K)dt)$$

Equilibrium distribution of the waiting time in a queue: Let T_n be the arrival time of the nth customer and let $T_{n+1}-T_n=U_{n+1}$. Let W_n be the waiting time for the nth customer and let V_n be his service time. Then, we have noted that

$$W_{n+1}=max(0,W_n+V_n-U_{n+1})=max(0,W_n-X_{n+1})$$

where

$$X_{n+1}=V_n-U_{n+1}$$

It follows that in the case of iid interarrival and service times with the two being being independent processes, X_n is an iid sequence and hence if $S_0=0$, then W_n has the same distribution as $max(S_k:0\leq k\leq n)$. So the equilibrium waiting time W_∞ has the same distribution as $sup(S_k:k\geq 0)$. Let τ_j denote the epoch of the jth ladder index for $j \geq 1$, ie, $\tau_j = n$ iff the process $S_k, k \geq 0$ attains its jth peak at time n and let $H_1 + ... + H_j = S_{\tau_j}$ denote the jth ladder height. Note that the process S_k is said to attain a peak at time n iff $S_n \geq max(S_k : k \leq n-1)$. Thus, $\tau_1 = n$ iff $max(S_k : k \leq n-1) \leq 0 < S_n$ and $H_1 = S_{\tau_1}$. In particular $(\tau_j, H_j), j = 1, 2, ...$ are iid bivariate random variables with $H_j = S_{\tau_1 + .. + \tau_j} - S_{\tau_1 + .. + \tau_{j-1}}, j \geq 1$ where $\tau_0 = 0$ by definition. Let L denote the distribution of H_1. Thus, L is concentrated over $[0, \infty)$ and

$$L(A) = \sum_{n \geq 1} Pr(max_{k \leq n-1} S_k \leq 0 < S_n \in A) = \sum_{n \geq 1} P(\tau_1 = n, S_n \in A)$$

Then, define

$$\psi_n(A) = \sum_{j \geq 1} P(\tau_1 + .. + \tau_j = n, S_n \in A), n \geq 1$$

Thus, $\psi_n(A)$ is the probability that a ladder epoch occurs at time n and that the corresponding ladder height is in A. We have

$$\sum_{n \geq 1} \psi_n(A) = \sum_{n,j \geq 1} P(\tau_1 + .. + \tau_j = n, S_n \in A)$$

$$= \sum_{n,j \geq 1} P(\tau_1 + ... + \tau_j = n, H_1 + ... + H_j \in A)$$

$$= \sum_{j \geq 1} P(H_1 + .. + H_j \in A) = \sum_{j \geq 1} L^{j*}(A)$$

Define $\psi_0(A) = 1$ if $0 \in A$ and $\psi_0(A) = 0$ if $0 \neq A$, ie, $\psi_0(A) = \delta_0(A)$. Then put

$$\psi(A) = 1 + \sum_{n \geq 1} \psi_n(A) = 1 + \sum_{j \geq 1} L^{j*}(A)$$

We have

$$\psi_n(A) = P(max_{k \leq n-1} S_k \leq S_n \in A) =$$
$$P(min_{k \leq n-1}(S_n - S_k) \geq 0, S_n \in A) =$$
$$P(min_{1 \leq k \leq n} S_k \geq 0, S_n \in A)$$

and further, if F denotes the probability distribution of X_1, then for $x \geq 0$,

$$P(min_{1 \leq k \leq n} S_k \geq 0, S_{n+1} \leq x) =$$

$$\int P(min_{1 \leq k \leq n} S_k \geq 0, S_n \in dy, X_{n+1} \leq x - y)$$

$$= \int_{y \geq 0} P(min_{1 \leq k \leq n} S_k \geq 0, S_n \in dy) F(x - y)$$

$$= \int_{y \geq 0} \psi_n(dy) F(x - y) = \int_{-\infty}^{x} \psi_n(x - y) dF(y)$$

Remark: Suppose that the dynamics of a stochastic system depends upon a parameter θ as for example queueing system in which the arrival rates and the service time rates are parameters which determine the probability distribution of the state at any given time t. Denoting by θ these parameters, the probability distribution at time t of the system state at time t is denoted by the vector $P(t|\theta)$. Suppose that this parameter θ undergoes a random fluctuation say $\delta\theta$. Then, we wish to control the non-random component θ of the parameter vector so that the probability distribution of the state is as close as possible to a given probability distribution $Q(t)$, ie we wish that the deviation probability $P(sup_{t\in][0,T]}|P(t|\theta+\delta\theta)-Q(t)| \geq \epsilon)$ to be as small as possible. Note that in this computation, we are computing the probability of deviation of a random probability distribution. This deviation probability can be computed approximately using the large deviation principle. In this formalism, the parameter θ can even be a function of time $\theta(t), t \in [0,T]$ and then its fluctuation $\delta\theta(t), t \in [0,T]$ will then be a random process. For example, we may write $\sqrt{\epsilon}\delta\theta(t)$ for the parameter fluctuation where $\delta\theta(.)$ is a Brownian motion process.

Chapter 6

LDP in Pattern Recognition and Fermionic Quantum Filtering

6.1 Large deviation problems in pattern recognition

Under the hypothesis H_1, X has the distribution $N(\mu_1, \Sigma)$ while under the hypothesis H_0, X has the distribution $N(\mu_0, \Sigma)$. Let P_1, P_0 denote the apriori probabilities of H_1 and H_0 occurring respectively. The optimal MAP test is then to select H_1 if $P(H_1|X) > P(H_0|X) and$ select H_0 otherwise. Now

$$P(H_1|X) = \frac{P(X|H_1)P_1}{P(X|H_1)P_1 + P(X|H_0)P_0}$$

$$= \frac{1}{1 + exp(-l(X))}$$

where

$$l(X) = ln(P(X|H_1)P_1/P(X|H_0)P_0))$$

$$= ln(P_1/P_0) + (1/2)((X-\mu_0)^T\Sigma^{-1}(X-\mu_0)$$

$$-(X-\mu_1)^T\Sigma^{-1}(X-\mu_1))$$

$$= ln(P_1/P_0) + (1/2)(\mu_0^T\Sigma^{-1}\mu_0 - \mu_1^T\Sigma^{-1}\mu_1)$$

$$+(\mu_1 - \mu_0)^T\Sigma^{-1}X$$

$$= W^TX + b$$

where

$$b = ln(P_1/P_0) + (1/2)(\mu_0^T\Sigma^{-1}\mu_0 - \mu_1^T\Sigma^{-1}\mu_1))$$

and

$$W = \Sigma^{-1}(\mu_1 - \mu_0)$$

Defining the sigmoidal function by

$$\sigma(x) = (1 + exp(-x))^{-1}, x \in \mathbb{R}$$

we see that $\sigma(x)$ increases from zero at $-\infty$ to 1 at $+\infty$ and that the optimal MAP test can be implemented as

$$Z_1 = \{X : \sigma(W^T X + b) > 1/2\}, Z_0 = \{X : \sigma(W^T X + b) \leq 1/2\}$$

Now we pose the following large deviation problem. Let $X_1, X_2, ...X_N$ be iid measurements taken either from the density $p(X|H_1)$ or from $p(X|H_0)$ and we have to make a decision. The MAP decision is to choose H_1 if

$$P(H_1|X_1, ..., X_N) = \frac{P(X_1, ..., X_N|H_1)P_1}{P(X_1, ..., X_N|H_1)P_1 + P(X_1, ..., X_N|H_0)P_0}$$

is greater than

$$P(H_0|X_1, ..., X_N) = \frac{P(X_1, ..., X_N|H_0)P_0}{P(X_1, ..., X_N|H_1)P_1 + P(X_1, ..., X_N|H_0)P_0}$$

and choose H_0 otherwise. Now

$$P(H_1|X_1, ..., X_N) = \frac{P_1 \Pi_{k=1}^N P(X_k|H_1)}{P_1 \Pi_{k=1}^N P(X_k|H_1) + P_0 \Pi_{k=1}^N P(X_k|H_0)}$$

$$= \frac{1}{1 + exp(-l(X_1, ..., X_N))}$$

where

$$l(X_1, ..., X_N) = ln(P_1/P_0) + \sum_{k=1}^{N} ln(P(X_k|H_1)/P(X_k|H_0))$$

Here we are not making any specific assumption about the distributions $P(X_k|H_r$ except that under each of the hypotheses, the r.v's are independent and identically distributed. Now as $N \to \infty$, we can apply Cramer's theorem to obtain the asymptotic distribution of

$$N^{-1} \sum_{k=1}^{N} ln(P(X_k|H_1)/P(X_k|H_0))$$

under each of the hypotheses. Letting $I_1(X)$ denote the rate function of this family of r.v's under H_1 and $I_0(X)$ the rate under H_0, We have for large N,

$$N^{-1}.ln(P(l(X_1, ..., X_N) \in B|H_1)) \approx -inf_{x \in B - ln(P_1/P_0)} I_1(x),$$

$$N^{-1}.ln(P(l(X_1, ..., X_N) \in B|H_0)) \approx -inf_{x \in B - ln(P_1/P_0)} I_0(x)$$

from which the asymptotic statistics of the sigmoidal function $\sigma(l(X_1, ..., X_N))$ under each of the hypotheses can be determined.

6.2 LDP for estimating the parameters in mixture models

Let $C_1, ..., C_p$ denote the p classes of the models. If X comes from C_k, then X has the density $p(X|\theta_k)$. The apriori probability of X coming from C_k is π_k so that $\sum_{k=1}^p \pi_k = 1$. If X comes from the mixture, then it has the density

$$p(X) = \sum_{k=1}^{p} \pi_k p(X|\theta_k)$$

Now a class C_k is selected at random in accordance with the apriori probabilities π_k and then a r.v. X is selected from that class. The class index is also noted. Thus, at the nth measurement, $t_n = k$ if C_k is the selected class and X_n is the measured r.v. We make N such independent measurements, noting at each measurement, both the class and the r.v. The sequence of measurements is thus the sequence of ordered pairs $\{(t_n, X_n) : n = 1, 2, ..., N\}$. The joint probability of getting the sequence $\{t_n\}$ and the density of the corresponding r.v's at X_n is thus given by

$$P((t_n, X_n) : n = 1, 2, ..., N|\theta_k, k = 1, 2, ..., N) =$$

$$\Pi_{n=1}^N \Pi_{k=1}^p (\pi_k P(X_n|\theta_k)^{\delta[t_n - k]})$$

We abbreviate this as $P((t, X)|\theta)$. Thus,

$$ln(P((t, X)|\theta)) = \sum_{n,p} \delta[t_n - k][ln(\pi_k) + ln(P(x_n|\theta_k))$$

and the MLE of $\theta = (\theta_k)$ may be obtained by maximizing this function.

6.3 The EM algorithm

Let $P(X|\theta)$ be the pdf of X and θ the parameter to be estimated. Maximizing this function is hard in general. However, there may be a latent r.v. Z such that maximizing $P(X|Z, \theta)$ or equivalently $P(X|Z, \theta)$ may be easier. We can write

$$ln(P(X|\theta)) = \int q(Z)ln(P(X, Z|\theta)/q(Z))dZ + \int q(Z).ln(q(Z)/P(Z|X, \theta))dZ$$

$$= T_1(q, X, \theta) + T_2(q, X, \theta)$$

say, for any probability density $q(Z)$. The first term on the rhs is non-positive and the second term is non-negative. Thus, we always have

$$ln(P(X|\theta)) \geq T_1(q, X, \theta) \forall q$$

The second term attains its minimum value of zero for a given X iff $q(Z) = P(Z|X,\theta)$ while the first term attains it maximum value of $ln(P(X|\theta))$ when $q(Z) = P(Z|X,\theta)$ because $P(X,Z|\theta) = P(Z|X,\theta).P(X|\theta)$.

Fix $\theta = \theta_1$ and then maximizing the first term T_1 on the rhs first w.r.t $q(.)$ gives $q(Z) = p(Z|X,\theta_1) = q_1$ say. This value of the first term on the rhs for this $q = q_1$ and general θ is then

$$T_1(q_1,X,\theta) = \int P(Z|X,\theta_1).ln(P(X,Z|\theta))dZ - \int P(Z|X,\theta_1).ln(P(X,Z|\theta_1))dZ$$

Note that

$$ln(P(X|\theta)) = T_1(q_1,X,\theta) + T_2(q_1,X,\theta) \geq T_1(q_1,X,\theta)$$

since

$$T_2(q_1,X,\theta_1) = 0 \leq T_2(q,X,\theta)\forall q,\theta$$

and further,

$$ln(P(X|\theta_1)) = T_1(q_1,X,\theta_1) + T_2(q_1,X,\theta_1) = T_1(q_1,X,\theta_1)$$

The next step is to calculate $\theta = \theta_2$ by maximizing $T_1(q_1,X,\theta)$ w.r.tθ, or equivalently, maximizing

$$\int P(Z|X,\theta_1).ln(P(X,Z|\theta))dZ$$

w.r.t θ. This amounts to maximizing $ln(P(X|\theta))$ provided that we assume that θ_1 is close to θ and hence that the positive quantity $T_2(q_1,X,\theta)$ can be neglected in the exact equation

$$ln(P(X|\theta)) = T_1(q_1,X,\theta) + T_2(q_1,X,\theta)$$

After that we set $q_2 = P(Z|X,\theta_2)$ and proceed in the same way. In other words,

$$\theta_{n+1} = argmax_\theta \int P(Z|X,\theta_n).ln(P(X,Z|\theta))dZ$$

Note:

$$ln(P(X|\theta_2)) = T_1(q_1,X,\theta_2) + T_2(q_1,X,\theta_2)$$
$$\geq T_1(q_1,X,\theta_1) + T_2(q_1,X,\theta_1) = ln(P(X|\theta_1))$$

because of the facts

$$T_1(q_1,X,\theta_2) \geq T_1(q_1,X,\theta_1),$$

and

$$T_2(q_1,X,\theta_2) \geq 0 = T_2(q_1,X,\theta_1)$$

where we use the definition of θ_2:

$$\theta_2 = argmax_\theta T_1(q_1,X,\theta)$$

The above discussion shows that it is reasonable to expect that θ_n converges to the ML estimate $\theta_{ML} = argmax_\theta ln(P(X|\theta))$ of θ. Note that we can write

$$\theta_{n+1} = argmax_\theta \int P(X,Z|\theta_n).ln(P(X,Z|\theta)/P(X,Z|\theta_n))dZ$$

The LDP problem: If $(X,Z)_N = \{(X_n,Z_n) : n = 1,2,...,N\}$ are iid with pdf $P(X,Z|\theta)$ and the EM method is used to estimate θ, then at what rate does the nth EM iterate $\theta_{N,n}$ based on $(X,Z)_N$ converge to a given r.v ? Further, is there an LDP for the family $\theta_{N,n}, N, n \geq 1$?

6.4 Sanov's theorem and Gibbs distributions

Let $X_n, n = 1,2,...$ be iid with distribution μ and let $L_N = N^{-1}\sum_{n=1}^N \delta_{X_n}$ be the empirical distribution of X based on the first N samples. Then, we know from Sanov's theorem that

$$N^{-1}.ln(P(L_N \in B)) \approx -inf(H(\nu|\mu) : \nu \in B), N \to \infty$$

where

$$H(\nu|\mu) = \int d\nu.ln(d\nu/d\mu)$$

Now we consider the Gibbs conditioning problem: For large N, calculate the most probable value of the probability distribution L_N conditioned on the event $\int f dL_N = E$, i.e $\int f(x)L_N(dx) = E$. This problem has the following physical interpretation: Given that there are N particles in the system located at iid random points $X_i, i = 1,2,...,N$ and given that a particle at x has a potential energy $f(x)$, it follows that the average energy of the system of particles is $N^{-1}\sum_{i=1}^N f(X_i)$. We are fixing this average energy at E and then we wish to find that under this constraint what is the most probable value of the empirical distribution of these particles as the number of particles becomes infinite. This most probable empirical distribution ν will be that which maximizes

$$P(L_N = \nu | \int f dL_N = E)$$

or equivalently that ν which maximizes

$$N^{-1}.lnP(L_N = \nu) - N^{-1}.ln(P(\int f dL_N = E)) - - - (1)$$

subject to the constraint $\int f dL_N = \nu$. As $N \to \infty$, by Sanov's theorem the lhs of (1) converges to

$$-H(\nu|\mu) + inf_{\rho \in B} H(\rho|\mu)$$

where

$$B = \{\nu : \int f d\nu = E\}$$

Thus, it becomes clear that the optimal value of ν as $N \to \infty$ is that probability distribution which minimizes $H(\nu|\mu)$ over all $\nu \in B$. Using Lagrange multipliers, this optimization problem reduces to minimizing

$$F(\nu, \lambda) = H(\nu|\mu) - \lambda.(\int f d\nu - E)$$

Suppose that we assume that ν has a density q w.r.t μ, ie,

$$q(x) = (d\nu/d\mu)(x)$$

Then the optimization problem is to minimize

$$F(q, \lambda_1, \lambda_2) = \int q(x).log(q(x))d\mu(x)$$
$$-\lambda_1(\int f(x)q(x)d\mu(x) - E) - \lambda_2(\int q(x)d\mu(x) - 1)$$

and from elementary variational calculus, the optimal q satisfies

$$log(q(x)) + 1 - \lambda_1 f(x) - \lambda_2 = 0, \int f(x)q(x)d\mu(x) = E, \int q(x)d\mu(x) = 1$$

or equivalently,

$$q(x) = C.exp(\lambda_1 f(x))$$

where C, λ_1 are determined from the conditions

$$C. \int exp(\lambda_1 x)d\mu(x) = 1, C. \int f(x).exp(\lambda_1 x)d\mu(x) = E$$

This is the classical Maxwell-Boltzmann-Gibbs distribution occurring in statistical mechanics wherein the optimal distribution of particles is obtained by maximizing the entropy subject to an energy constraint.

6.5 Gibbs distribution in the interacting particle case

So far, we have been considering non-interacting particles, ie, the case when the total potential energy of the system of particles is a sum of the individual energies. Suppose now that the total potential energy has interaction terms:

$$E(N) = N^{-2} \sum_{1 \leq i,j \leq N} U(X_i, X_j), U(x, y) = U(y, x)$$

where $X_1, ..., X_N$ are as before iid random variables with probability distribution μ. Consider

$$\int U(x, y)dL_N(x)dL_N(y) = N^{-2} \sum_{i,j=1}^{N} U(X_i, X_j) = E(N)$$

and hence as $N \to \infty$, the problem of determining the optimal probability distribution ν for which $P(L_N = \nu | \int U(x,y)dL_N(x)dL_N(y) = E)$ is a maximum amounts to minimizing

$$H(\nu|\mu)$$

subject to the constraint

$$\int U(x,y)d\nu(x)d\nu(y) = E$$

and in terms of the density $q(x) = (d\nu/d\mu)(x)$, the function to be minimized is

$$F(q,\lambda_1,\lambda_2) = \int q(x).ln(q(x))d\mu(x) - \lambda_1(\int U(x,y)q(x)q(y)d\mu(x)d\mu(y) - E)$$

$$-\lambda_2(\int q(x)d\mu(x) - 1)$$

3.6 Inversion of the characteristic function of a probability distribution on the real line

Let F be a probability distribution on $\mathbb{R}$ and let $\phi(t)$ be its characteristic function:

$$\phi(t) = \int_{\mathbb{R}} exp(itx)dF(x), t \in \mathbb{R}$$

Then, by Fubini's theorem,

$$\int_{-T}^{T}\int_{a}^{b} phi(t)exp(-itx)dtdx$$

$$= \int_{-T}^{T}(exp(-ita) - exp(-itb))\phi(t)dt/it = \int_{\mathbb{R}} dF(y)\int_{-T}^{T}\int_{a}^{b} exp(it(y-x))dtdx$$

$$= 2\int_{\mathbb{R}} dF(y)\int_{a}^{b} sin(T(y-x))dx/(y-x) - - - (1)$$

Now,

$$\int_{a}^{b} sin(T(y-x))dx/(y-x) = \int_{T(a-y)}^{(T(b-y)} sin(u)du/u - - - (2)$$

For $a < y < b$, this converges to π, for $y = a$, it converges to $\pi/2$, for $y = b$, it converges to $\pi/2$ and for $y > b$ or $y < a$, it converges to zero as $T \to \infty$. This is because

$$\int_{0}^{\infty} sin(u)du/u = \pi/2$$

and hence

$$\int_{-\infty}^{\infty} sin(u)du/u = \pi$$

Thus application of the bounded convergence theorem to (1) gives

$$(2\pi)^{-1}\int_{-\infty}^{\infty}\phi(t)(exp(-ita)-exp(-itb))dt/it=$$

$$F(b-)-F(a)+(F(b)-F(b-)/2=(F(b)+F(b-))/2-F(a)$$

on making use of the right continuity of F.

6.7 Infinitely divisible distributions, The Levy Khintchine theorem

Lemma: Let $z_k, w_k, k = 1,2,...,N$ be complex numbers bounded by one i magnitude. Then,

$$|\Pi_{k=1}^{N} z_k - \Pi_{k=1}^{N} w_k| \leq \sum_{k=1}^{N} |z_k - w_k|$$

6.8 Stationary distribution for Markov chains

Let $P = ((P(i,j))$ be a stochastic matrix in $\mathbb{R}^{\mathbb{Z}_+\times\mathbb{Z}_+}$. Assume that is its irre ducible and aperiodic. Let X_n and Y_n be two independent Markov chains bot having P as their one step transition probabilities. Suppose P is a transien chain. Then it cannot have a stationary distribution, for if π is a stationar distribution, then

$$\sum_i \pi(i)P(i,j)=\pi(j)$$

which implies

$$\sum_i \pi(i)P^{(n)}(i,j)=\pi(j), n=1,2,...--(1)($$

and the transience of the chain implies that $\sum_n P^{(n)}(i,j)<\infty$ for all i,j an hence in particular,

$$P^{(n)}(i,j)\rightarrow 0, n\rightarrow\infty\forall i,j$$

Taking the limit on both sides of (1) and using the dominated convergenc principle then yields

$$\pi(j)=lim_{n\rightarrow\infty}\sum_i \pi(i)P^{(n)}(i,j)=0\forall j$$

which means that π is zero and hence P cannot have any stationary distributior On the other hand, suppose π is a stationary distribution. Then, P is a persis tent chain, ie $\sum_n P^{(n)}(i,j)=\infty\forall i,j$. Then the coupled chain $(X_n,Y_n), n\geq$

is also persistent with stochastic matrix $P((i,j),(k,l)) = P(i,k)P(j,l)$. Indeed, $\pi_i\pi_j$ is now a stationary distribution for the coupled chain and hence the coupled chain must be persistent by the same logic as used above for the original chain.

Let τ be the first time at which the coupled chain visits (i_0, i_0). For $n \geq m$,

$$P_{(i,j)}((X_n, Y_n) = (k,l), \tau = m) =$$

$$= P_{i,j)}(\tau = m)P_{(i_0,i_0)}((X_{n-m}, Y_{n-m}) = (k,l))$$

$$= P_{(i,j)}(\tau = m)P^{(n-m)}(i_0, k)P^{(n-m)}(i_0, l)$$

Summing over l gives

$$P_i(X_n = k, \tau = m) = P_{(i,j)}(\tau = m)P_{i_0}(X_{n-m} = k)$$

and summing over k gives

$$P_j(Y_n = l, \tau = m) = P_{(i,j)}(\tau = m)P_{i_0}(Y_{n-m} = l)$$

Replacing l by k in this last equation gives

$$P_j(Y_n = k, \tau = m) = P_{(i,j)}(\tau = m)P_{i_0}(Y_{n-m} = k)$$

Noting that X_n and Y_n are identically distributed processes, we get

$$P_i(X_n = k, \tau = m) = P_j(Y_n = k, \tau = m), n \geq m$$

Then, summing over $m = 0, 1, ..., n$ gives us

$$P_i(X_n = k, \tau \leq n) = P_j(Y_n = k, \tau \leq n)$$

and hence

$$|P_i(X_n = k) - P_j(Y_n = k)| = |P(X_n = k, \tau > n) - P(Y_n = k, \tau > n)|$$

$$\leq 2P(\tau > n)$$

and hence letting $n \to \infty$ and noting that the persistence of the coupled chain implies $P(\tau > n) \to 0$, we get

$$lim_{n\to\infty}|P^{(n)}(i,k) - P^{(n)}(j,k)| = 0$$

for all states i, j, k. This result implies that if $limP^{(n)}(i,k)$ exists, then this limit is independent of i and hence, it follows from the equation

$$\pi(k) - \sum_j \pi(j)P^{(n)}(j,k)$$

for a stationary distribution π that if $lim_n P^{(n)}(i,k)$ exists then this limit is independent of i and in this case, if a stationary distribution π exists, then it is unique and is given by

$$\pi(k) = lim_n P^{(n)}(i,k)$$

which does not depend on the state i.

Remark 1: If P is an irreducible chain, then all its states have the same period. If in addition, it is aperiodic, then since

$$P^{(n+m)}(i,i) \geq P^{(n)}(i,i).P^{(m)}(i,i), n, m \geq 0 \forall i$$

it follows that the set

$$D_i = \{n : P^{(n)}(i,i) > 0\}$$

is closed under addition and since D_i has in addition $gcd = 1$ (by definition of aperidoicity), it follows that there exists an integer $n_0(i)$ such that every $n > n_0(i)$ is in D_i.

Remark 2: Let D be a set of positive integers closed under addition and having $gcd = 1$. Then D contains all integers greater than some finite positive integer. To see this, first observe that since D is closed under addition, D is an infinite set. Let then $n(k), k = 1, 2, ...$ is an increasing sequence of positive integers. Since $gcd(D) = 1$, it follows that there is a finite positive integer K such that $gcd(n(k) : k \leq K) = 1$. Hence, there exist integers $p(1), .., p(K)$ such that

$$\sum_{k=1}^{K} p(k)n(k) = 1$$

This equation can be expressed as

$$\sum_{k \in I} p(k)n(k) = \sum_{k \in I^c} q(k)n(k) + 1$$

where I is a subset of $\{1, 2, ..., K\}$ and I^c is the complement of I in $\{1, 2, ..., K\}$. Here $p(k), q(j)$ are non-negative integers. It follows that since D is closed under addition that

$$\sum_{k \in I} p(k)n(k), \sum_{k \in I^c} q(k)n(k) \in D$$

In other words, we have proved the existence of an element $r \in D$ such that $r + 1 \in D$. Note that r is a positive integer. Now let n be any positive integer. We have by the Euclidean division algorithm that

$$n = qr + s$$

where $0 \leq s \leq r-1$ and if $n \geq r^2$, then $q \geq r > s$. Hence we can write $q = s+m$ where m is a positive integer. Thus,

$$n = (s + m)r + s = s(r + 1) + mr$$

and from this expression, since $r, r + 1 \in D$ and D is closed under addition, it follows that $n \in D$.

6.9 Lecture on quantum filtering in the presence of Fermionic noise

Aim of this lecture: Suppose we have a cavity resonator system with the environment being a noisy quantum bath. The cavity system contains N oscillating photon modes along with M Fermions (electrons and positrons). The system can therefore be described by a Hilbert space $L^2(\mathbb{R})^{\otimes_s N} = \mathcal{H}_b^{\otimes_s N}$ that supports the N photons, tensored with another Hilbert space $\mathcal{H}_f^{\otimes_a M}$ that supports the M Fermions. Here $\otimes_s$ denotes symmetric tensor product while $\otimes_a$ denotes antisymmetric tensor product. The creation and annihilation operators of the photons are denoted by $c(k)^*, c(k), k = 1, 2, ..., N$ while those of the Fermions are $a(k)^*, a(k), k = 1, 2, ..., M$. These operators satisfy the canonical bosonic commutation and Fermionic anticommutation relations:

$$[c(k), c(m)^*] = \delta[k-m], [a(k), a(m)^*]_+ = \delta[k-m],$$

$$[c(k), c(m)] = [c(k)^*, c(m)^*] = 0, [a(k), a(m)]_+ = [a(k)^*, a(m)^*]_+ = 0$$

and any bosonic operator commutes with any Fermionic operator. The system Hilbert space is therefore

$$\mathfrak{h} = \mathcal{H}_b^{\otimes_s N} \otimes \mathcal{H}_f^{\otimes_a M}$$

Note that if $dim\mathcal{H}_f = p$, then M can be atmost p. Now the Hilbert space $\mathfrak{h}$ has a natural $\mathbb{Z}_2$ grading. Any vector $v \in \mathfrak{h}$ which is of the form $u \otimes v$ where $u \in \mathcal{H}_b^{\otimes_s N}$ and v is a superposition of vectors of the form $v_1 \otimes_a ... \otimes_a v_r$ with r even is said to be of even parity while any vector which is of the same from with r odd is said to be of odd parity. Thus, we can express the system Hilbert space as the direct sum of an even Hilbert space and an odd Hilbert space:

$$\mathfrak{h} = \mathfrak{h}_0 \oplus \mathfrak{h}_1$$

where $\mathfrak{h}_0$ is spanned by vectors of the from $u \otimes v$ with $u \in \mathcal{H}_b^{\otimes_s N}$ and $v \in \mathcal{H}_a^{\otimes M}$ with v being the antisymmetric tensor product of an even number of vectors while $\mathfrak{h}_1$ is the same but with v being the antisymmetric tensor product of an odd number of vectors. We can also consider the infinite dimensional setting in which there are an infinite number of photons and Fermions. In that case, the bosonic Hilbert space is a Boson Fock space with respect to a given Hilbert space $\mathcal{H}_b$ while the Fermionic Hilbert space is a Fermionic Fock space with respect to another Hilbert space $\mathcal{H}_f$. The system Hilbert space is therefore

$$\mathfrak{h} = \Gamma_s(\mathcal{H}_b) \otimes \Gamma_a(\mathcal{H}_f)$$

where

$$\Gamma_s(\mathcal{H}_b) = \mathbb{C} \oplus \bigoplus_{n \geq 1} \mathcal{H}_b^{\otimes_s n},$$

$$\Gamma_a(\mathcal{H}_f) = \mathbb{C} \oplus \bigoplus_{n \geq 1} \mathcal{H}_f^{\otimes_a n}$$

Again, $\mathfrak{h}$ has the natural structure of a $\mathbb{Z}_2$ graded super-Hilbert space.

$$\mathfrak{h} = \mathfrak{h}_0 \oplus \mathfrak{h}_1$$

A typical physical situation in which such a system arises is quantum electrodynamics in which the photon field is the second quantized four vector potential:

$$A^\mu(x) = \int [(c(k,s)/\sqrt{2|K|})e^\mu(K,s)exp(-ik.x)+(c(k,s)^*/\sqrt{2|K|})\bar{e}^\mu(K,s)exp(ik.x)]d^3K$$

with $k^0 = |K|$ and the Fermionic field is the Dirac electron-positron field described by the four component wave operator field

$$\psi(x) = \int [u(P,\sigma)a(P,\sigma)exp(-ip.x) + v(P,\sigma)b(P,\sigma)^* exp(ip.x)]d^3P$$

with $u(P,\sigma), v(-P,\sigma)$ are the eigenvectors of the Dirac Hamiltonian in the momentum domain with eigenvalues $p^0 = \sqrt{P^2+m^2}$ and $-p^0 = -\sqrt{P^2+m^2}$. σ takes two values $\pm 1/2$ corresponding to the fact that the electrons and positrons are spin $1/2$ particles. $a(P,\sigma), b(P,\sigma)^*$ are respectively the electron annihilation and positron creation operators in the momentum-spin domain. This structure of the cavity field is obtained using the Lagrangian density of the free electromagnetic field

$$L_E(A_\mu, A_{\mu,\nu}) = (-1/4)F_{\mu\nu}F^{\mu\nu}, F_{\mu\nu} = A_{\nu,\mu} - A_{\mu,\nu}$$

and that of the free Dirac field

$$L_D(\psi, \bar{\psi}, \partial_\mu\psi) = \bar{\psi}(i\gamma^\mu\partial_\mu - m)\psi$$

The equations of motion for the free electromagnetic field and that of the free Dirac field are derived from the Euler-Lagrange equations and these are respectively the wave equation for the four vector potential and the Dirac equation for the four component wave function. Solving these free wave equations gives us the above expansions of the field in terms of the field creation and annihilation operators. These Lagrangians also take as position fields A_μ, ψ respectively and identify the corresponding momentum fields as $\partial L_E/\partial A_{\mu,0}$ and $\partial L_D/\partial\psi_{,0}$ and then we set up the canonical commutation and canonical anticommutation relations between these canonical position and momentum fields and these translate into the CCR and CAR for the field creation and annihilation operators:

$$[c(K,s), c(K',s')^*] = \delta^3(K-K')\delta_{s,s'},$$

with all the other bosonic commutation relations being zero,

$$[a(P,\sigma), a(P',\sigma')^*]_+ = \delta^3(P-P')\delta_{\sigma,\sigma'}$$

$$[b(P,\sigma), b(P,\sigma')^*]_+ = \delta^3(P-P')\delta_{\sigma,\sigma'}$$

with all the other Fermionic anticommutation relations being zero. On discretization in the momentum domain, we get the model mentioned at the beginning with $(c[k], c[k]^*)$ being discretized versions of $(c(K,s), c(K,s)^*)$, and $(a(k), a(k)^*)$ being discretized versions of $(a(P,\sigma), a(P,\sigma)^*)$ and $(b(P,\sigma), b(P,\sigma)^*)$. We can also consider the interaction energy between the Dirac field and the photon field, ie, $H_I(t) = \int J^\mu(x)A_\mu(x)d^3x$ where $J^\mu(x) = -e\bar{\psi}(x)\gamma^\mu\psi(x)$ is the Dirac four current density. The total Hamiltonian of the photon and Fermion fields in the second quantized picture is then given by

$$H(t) = H_{ph} + H_f + H_I(t)$$

where H_{ph}, the Hamiltonian of the photon field is given by

$$H_{ph} = (1/2)\int (E^2 + B^2)d^3x, E_r = F_{0r}, B_r = \epsilon(rkm)F_{km}$$

H_f, the Hamiltonian of the fermionic field is given by

$$H_f = \int \psi(x)^*((\alpha, -i\nabla) + \beta m)\psi(x)d^3x$$

and $H_I(t)$, the interaction Hamiltonian between the photon and the fermion field is given by

$$H_I(t) = -e\int \bar{\psi}(x)\gamma^\mu\psi(x)A_\mu(x)d^3x$$

It is easy to show by substituting for the photon and Dirac fields A_μ, ψ, their expressions in terms of the photon and fermion creation and annihilation operators that

$$H_{ph} = \int |K|c(K,s)^*c(K,s)d^3K - - - (1)$$

which on discretization becomes

$$H_{ph} = \sum_{k=1}^{N} \omega(k)c(k)^*c(k) - - - (2)$$

The Fermion Hamiltonian is

$$H_f = \int E(P)(a(P,\sigma)^*a(P,\sigma) + b(P,\sigma)^*b(P,\sigma))d^3P$$

which on discretization becomes

$$H_f = \sum_{k=1}^{M} E(k)a(k)^*a(k)$$

Generally for confined fields, ie, for fields encosed in a cavity, the continuous integral (1) will get discretized as also will the frequencies $c|K|$ get replaced by the countable set of characteristic frequencies of oscillation of the cavity modes.

This discretization of the characteristic frequencies arises owing to the appl caition of the boundary conditions on the boundary walls of the cavity. Thu in (2), the $\omega(k)'s$ are regarded as the characteristic frequencies of oscillation c the photon modes. Finally, the interaction Hamiltonian has the form

$$H_I(t) = \sum_{k,m,r} (C_1(t,k,m,r)a(k)^*a(m)+$$

$$C_2(t,k,m,r)a(k)a(m)+C_3(t,k,m,r)a(k)^*a(m)^*)c(r)+H.C$$

In other words, $H_I(t)$ is a homogeneous cubic polynomial in the boson-fermio creation-annihilation operators in such a way that is it is linear in the photo creation-annihilation operators and quadratic in the fermion creation-annihilati operators. It is $H_I(t)$ that is responsible for the interaction processes that tak place between the electrons, positrons and photons like scattering, absorptior emission giving rise to effects such as Compton scatterin in which an electro absorbs a photon, moves forward and then remits a photon of perhaps a differ ent frequency, or vacuum polarization in which a photon disintegrates into a electron-positron pair which then recombine by annihilating each other thereb once again generating a photon. Some of these processes are responsible fc radiative effects like the electron self energy, the anomalous magnetic momen of the electron etc which in turn can be calculated by considering Feynma diagrams with loops inserted in propagator lines, the loops corresponding t vacuum polarization or other effects. A splendid review of these effects can b found in [1].

The operators $c(k), c(k)^*$ are even system operators while the operator $a(k), a(k)^*$ are odd system operators. A monomial in these operators havin an even number of the latter operators is even while that having an odd numbe of these operators is odd. To describe this situation better we observe that i the orthogonal decomposition

$$\mathfrak{h} = \mathfrak{h}_0 \oplus \mathfrak{h}_1$$

we let P_0 denote the projection onto $\mathfrak{h}_0$ so that $P_1 = 1 - P_0$ becomes th projection onto $\mathfrak{h}_1$. We define the grading operator in $\mathfrak{h}$ by

$$\theta = P_0 - P_1$$

An operator X in $\mathfrak{h}$ is said to be even if

$$X(\mathfrak{h}_k) \subset \mathfrak{h}_k, k = 0, 1$$

and odd if

$$X(\mathfrak{h}_0) \subset \mathfrak{h}_1, X(\mathfrak{h}_1) \subset \mathfrak{h}_0$$

It is clear that X is even iff

$$X = P_0XP_0 + P_1XP_1$$

and X is odd iff

$$X = P_0XP_1 + P_1XP_0$$

It is also clear that X is even iff

$$P_0XP_1 + P_1XP_0 = 0$$

iff

$$\theta X\theta = X$$

while X is odd iff

$$P_0XP_0 + P_1XP_1 = 0$$

iff

$$\theta X\theta = -X$$

We write

$$\tau(X) = \theta.X.\theta$$

Thus, X is even iff $\tau(X) = X$ and is odd iff $\tau(X) = -X$. Given any operator X in $\mathfrak{h}$, it can be expressed in a unique way as

$$X = X_+ + X_-$$

where X_+ is even and X_- is odd. In fact,

$$X = (P_0+P_1)X(P_0+P_1) = (P_0XP_0+P_1XP_1)+(P_0XP_1+P_1XP_0) = X_+ + X_-$$

where

$$X_+ = P_0XP_0 + P_+1XP_1$$

is even and

$$X_- = P_0XP_1 + P_1XP_0$$

is odd. For anhy operator X,

$$X = X_+ + X_-, \tau(X) = X_+ - X_-$$

Now given two Hilbert spaces $\langle_k, k = 1, 2$, with tensor product $\otimes$ between them, we define a graded tensor product $\otimes_g$ between two operators X, Y in these spaces as

$$X \otimes_g Y = X\theta_1 \times Y$$

where θ_1 is the grading operator in $\mathfrak{h}_1$. Then, we have

$$(X_1 \times_g Y_1).(X_2 \otimes_g Y_2) = X_1\theta_1 X_2\theta_1 \otimes Y_1Y_2$$

$$= X_1\tau_1(X_2) \otimes Y_1Y_2$$

In particular,

$$(X \otimes_g I_2)(I_1 \otimes_g Y) = X \otimes Y, (I_1 \otimes_g Y)(X \otimes_g I_2) = \tau_1(X) \otimes Y$$

Hence if X is even, then

$$(X \otimes_g I_2)(I_1 \otimes_g Y) = (I_1 \otimes_g Y)(X \otimes_g I_2)$$

while if X is odd, then

$$(X \otimes_g I_2)(I_1 \otimes_g Y) = -(I_1 \otimes_g Y)(X \otimes_g I_2)$$

In the context of system and bath in quantum stochastic calculus theory, an even system operator commutes with bath noise operators while an odd system operator anticommutes with bath noise operators provided that we use everywhere the graded tensor product between system and bath Hilbert spaces. In our cavity filtering problem, the system operators that modulate the fermionic bath noise operator processes are odd, so that these two anticommute in accordance with the above formalism. However, for filtering in the presence of Fermionic noise to work out, we require the measurement process to be the bath quantum Poisson passed through the system, ie the measurement process should be of the photon counting type, rather than fermionic Brownian motion because otherwise, we would not get the non-demolition property satisfied. Now Let $J(t)$ be fermionic Brownian motion and let L_1, L_2 be odd system operators. The qsde governing the unitary dynamics of system and bath evolution is given by

$$dU(t) = (-(iH(t) + P)dt + L_1 dJ(t) - L_2 dJ(t)^* + Sd\Lambda(t))U(t)$$

The Fermionic quantum Ito formula

$$dJ(t).dJ(t)^* = dt, [J(t), J(s)^*]_+ = min(t,s), [J(t), J(s)]_+ = 0 = [J(t)^*, J(s)^*]_+$$

and of course

$$dJ(t)d\Lambda(t) = dJ(t), d\Lambda(t)dJ(t)^* = dJ(t)^*, (d\Lambda(t))^2 = d\Lambda(t)$$

Note that by the property of odd system operators (as discussed above in terms of the graded tensor product)

$$L_1 dJ(t) = -dJ(t)L_1, L_2 dJ(t)^* = -dJ(t)^* L_2$$

Note that S is an even system operator and the condition for $U(t)$ to be unitary can be deduced by applying the quantum Ito formula to

$$0 = d(U(t)^*U(t)) = dU(t)^*.U(t) + U(t)^*.dU(t) + dU(t)^*.dU(t)$$

to get

$$-2P + L_2^* L_2 = 0, S^*S + S + S^* = 0, L_2^* S + L_1 + L_2^* = 0$$

Note that we've used the following identities

$$(L_2 dJ(t)^*)^* = dJ(t)L_2^* = -L_2^* dJ(t),$$

$$(L_2 dJ(t))^* Sd\Lambda(t) = -L_2^* dJ(t)Sd\Lambda(t) = -L_2^* SdJ(t)d\Lambda(t) = -L_2^* SdJ(t)$$

$$(L_1 dJ(t))^* = -dJ(t)^* L_1^* = L_1^* dJ(t)^*$$

$$(L_2 dJ(t)^*)^*(L_2 dJ(t)) = dJ(t)L_2^* L_2 dJ(t) = L_2^* L_2 dt$$

since $dJ(t)$ anticommutes with both L_2, L_2^* and hence commutes with $L_2^* L_2$. Note that $U(t)$ is an even operator and that L_1, L_2, S are functions of the system creation and annihilation operators of the photons and the Fermions chosen in such a way as explained above so that L_1, L_2 (and hence also their adjoints) are odd while S is even. Now consider the input measurement

$$Y_i(t) = \Lambda(t)$$

and the corresponding output photon counting measurement

$$Y_o(t) = U(t)^* Y_i(t) U(t) = U(t)^* \Lambda(t).U(t)$$

Note that $\Lambda(t)$ commutes with the system operators L_1, L_2 since the former is bosonic and hence even and we use the graded tensor product. Also $\Lambda(t)$ commutes with $dJ(T), dJ(T)^*, T > t$ which is proved using the usual representation $dJ(T) = (-1)^{\Lambda(T)} dA(T), dJ(T)^* = (-1)^{\Lambda(T)} dA(T)^*$. Thus $\Lambda(t)$ commutes with the combinations $L_1 dJ(T), L_2 dJ(T)^*, T > t$ and their adjoints which is what by virtue of the unitarity of $U(t)$ makes the output photon measurement satisfy

$$Y_o(t) = U(T)^* \Lambda(t) U(T), T \geq t$$

and hence guarantee that $Y_o(.)$ will be a non-demolition measurement, ie, if X is any system space observable, then $Y_o(t) = U(T)^* \Lambda(t).U(T)$ will commute with $j_T(X) = U(T)^* X U(T) \forall T \geq t$ since $\Lambda(t)$ commutes with X. Note that $J(t)$ and $J(t)^*$ anticommute with L_1, L_2 and their adjoints since $J(t), J(t)^*, L_1, L_2, L_1^*, L_2^*$ are all odd provided that we use the graded tensor product and also $J(t)$ anticommutes with $dJ(T) = (-1)^{\Lambda(T)} dA(T)$ for $T > t$. Thus, $J(t)$ also commutes with the combinations $L_1 dJ(T), L_2 dJ(T)^*, T \geq t$. The same holds for $J(t)^*$. This means that

$$\begin{aligned} Z_o(t) &= U(T)^* (c_1 \Lambda(t) + c_2 J(t) + \bar{c}_2 J(t)^*) U(T) \\ &= U(t)(c_1 \Lambda(t) + c_2 J(t) + \bar{c}_2 J(t)^*) U(t), T \geq t \end{aligned}$$

commutes with $j_T(X)$ for any system operator X. However $Z_o(.)$ cannot be used as a non-demolition measurement because for $t > s, c_2 dJ(t) + \bar{c}_2 dJ(t)^*$ does not commute with $c_2 dJ(s) + \bar{c}_2 dJ(s)^*$, rather, they anticommute. So to obtain a Bosonic-Fermionic mixed non-demolition measurement comprising of a Fermionic Brownian motion and Quantum Poisson process, we must allow $c_2, \bar{c}_2$ to be anticommuting Grassmannian parameters which commute with everything else. Then $Z_o(.)$ will be a non demolition Abelian family of measurements and the filtering theory will carry through for this. More generally, we can take

$$dZ_i(t) = c_1(t) d\Lambda(t) + c_2(t) dJ(t) + \bar{c}_2(t) dJ(t)^*, Z_o(t) = U(t)^* Z_i(t) U(t)$$

where $c_1(t)$ is a real (bosonic) valued function of time and $c_2(t), \bar{c}_2(t)$ are Grassmannian (fermionic) parameters which commute with everything else and mutually anticommute, ie,

$$c_2(t) c_2(s) + c_2(s) c_2(t) = 0, c_2(t) \bar{c}_2(s) + \bar{c}_2(s) c_2(t) = 0 \forall t, s,$$

Note that the first implies

$$\bar{c}_2(t)\bar{c}_2(s) + \bar{c}_2(s)\bar{c}_2(t) = 0$$

Then $Z_o(.)$ forms an Abelian family of non-demolition measurements and can be used to develop our mixed boson-fermion filter. Now let X be any system space observable. Then,

$$dj_t(X) = U(t)^*((iH - P)dt + dJ(t)L_1^* + dJ(t)^*L_2^* + S^*d\Lambda(t))XU(t)$$

$$+U(t)^*(-(iH + P)dt + L_2dJ(t) + L_2dJ(t)^* + Sd\Lambda(t))XU(t)$$

$$+U(t)^*(dJ(t)^*L_1^* + dJ(t)L_2^*)X(L_1dJ(t) + L_2dJ(t)^*)U(t)$$

Now,

$$dJ(t)L_2^*XL_2dJ(t)^* = \tau(L_2^*XL_2)dJ(t).dJ(t)^* = L_2^*\tau(X)L_2dt$$

Remark: We can also add bosonic noise differentials of the form $L_3dA(t) + L_4dA(t)^*$ into our evolution equation for $U(t)$ and readjust correspondingly the quantum Ito correction term P to make $U(t)$ unitary. L_3, L_4 would be bosonic, ie even system operators. Then, for example $d\Lambda(t)$ would commute with $L_3dA(T) + L_4dA(T)^*$ for $T \geq t$ and $dJ(t)$ would commute with $L_3dA(T) + L_4dA(T)^*$ for $T \geq t$ because $dJ(t)$ commutes with L_3, L_4, the latter being even and also commutes with $dA(T), dA(T)^*$ for $T \geq t$. The same is true for $dJ(t)^*$. We could also add bosonic noise terms into our measurement process $Z_i(t)$, these extra differentials contributing to $dZ_i(t)$ would be $c_3(t)dA(t) + c_4(t)dA(t)^*$. Note that $dA(t)$ anticommutes with L_1, L_2 and their adjoints, the latter being odd (relative to the graded tensor product) and it also anticommutes with $dJ(T), dJ(T)^*$ since it anticommutes with $(-1)^{\Lambda(T)}$ for $T \geq t$. Therefore, $dA(t)$ commutes with $L_1dJ(T), L_2dJ(T)^*$ for $T \geq t$. Thus, with these extra terms added to Z_i our measurement process Z_o becomes non-demolition, ie, it commutes with the future state values: $[Z_o(t), j_T(X)] = 0, T \geq t$. However, to make it Abelian, we note that $dA(t)$ anticommutes with $dJ(T), dJ(T)^*$ for $T \geq t$, so we require that $dA(t)$ also anticommute with the Grassmannian parameters $c_2(T), \bar{c}_2(T), T \geq t$ in order that it commute with $c_2(T)dJ(T) + \bar{c}_2(T)dJ(T)^*, T \geq t$.

5.10 Lecture plan for Pattern Recognition

1] Basic definition and examples of pattern recognition taken from speech, images and classical and quantum physics. Removing chaos and creating order.

[2] [a] Design principles of pattern recognition systems based on identification of non-random parameters from random data, training the weights of a neural network to match input-output data of a system. Examples of fuzzy neural networks.

[b] Examples of pattern recognition from image processing based on invariants of the image field under a Lie group of transformations.

[3] Learning and adaptation:

[a] Adapting the parameters of our model/weights of a neural network using the gradient search algorithm.

[b] Adaptation based on recursive least squares algorithm with forgetting factor.

[4] Pattern recognition approaches. [a] Parametric Methods for parameter estimation like ML, MAP, MMSE, LSE, WLSE, applied to models that are linear and nonlinear in the parameters like linear and nonlinear time series analysis.

[b] Parametric models based on modeling the probability distribution of the measured data using well known probability distributions with unknown parameters like mean, covariance and higher order moments.

[c] Non-parametric density estimation, non-parametric spectral and higher order spectral estimation as examples of non-parametric pattern recognition.

[5] Mathematical foundations of pattern recognition: Linear algebra, probability theory:

Linear models, sequential least squares estimation in linear models, least squares using generalized inverses in terms of the singular value decomposition, estimating parameters of the multivariate normal distribution, chi-square distribution.

[6] Statistical pattern recognition:

[a] Bayesian decision theory, classifiers, discriminant function using the likelihood ratio test.

[b] Quantum binary hypothesis testing: Discriminating between the two states of a quantum mechanical system using positive operator valued measures (POVM), the Holevo-Helstrom theory.

[c] A comparison between M-ary classical hypothesis testing and M-ary quantum hypothesis testing. Solving the M-ary quantum hypothesis testing problem using optimization algorithms.

[7] Maximum likelihood and Bayesian parameter estimation, basic theory with examples taken from classical mechanics, quantum mechanics, fluid dynamics (fluid velocity field parameter estimation), and classical and quantum field theory.

[8] Pattern recognition examples taken from statistical image processing.

[a] Examples of parameter estimation in Gaussian mixture models.

[b] Examples of parameter estimation taken from the problem of calculating

the group transformation element applied to an image field from noisy trans formed data.

[c] Some background in group representation theory.

Lie group, compact Lie group, permutation group, representation of a grou in a vector space, the Schur Lemmas and the Peter-Weyl theorem for com pact groups, Group theoretic Fourier transform, Representations of the rotatio and Lorentz group, induced representations, representations of the Galilean an Poincare group using induced representations of the semidirect product.

[9] Examples of pattern recognition taken from noisy speech and text dat based on real time nonlinear filtering algorithms.

[10] Performance analysis of statistical parameter estimators based on per turbation theory and large deviation principles.

[11] Dimension reduction and principal component analysis of random data

[12] Hidden Markov models.

[a] Markov chains and processes, general theory.

[b] Hidden Markov models:Estimating the transition probabilities from emis sion observations.

[c] Examples of HMM from speech and image sequence modeling.

[13] The Expectation Maximization (EM) algorithm. Intuitive proof of con vergence of the EM estimate to the ML estimate based on relative entropy.

[14] Relative entropy from the large deviation standpoint.

[a] Sanov's theorem on the asymptotic distribution of the empirical density

[b]

6.11 Review of the book Stochastics, control an robotics, by Harish Parthasarathy

The book studies various aspects of control of robots both classical and quantur using electromagnetic fields and other methods when the robot dynamics i subject to classical and quantum stochastic noise perturbations. Some of th important problems studied here are as follows:

A robot system in motion comprising 3-D links with each link carrying a cur rent density field radiates out into space and the problem of estimating the lin 3-D rotation matrices from measurement of these radiated fields is simplified us ing group representation theoretic Fourier transforms. Such a non-commutativ Fourier transform simplifies the problem of configuration estimation into a linea problem.

Quantum filtering as developed by Belavkin is discussed from the viewpoin of estimating functions of the quantum robot angular position and velocity (ie Heisenberg observables) or for estimating the evolving quantum state of th robot (Schrodinger picture) from non-demolition measurements. It is impor tant to take such measurements to determine the robot state/observables sinc demolition measurements prevent the possibility of such an estimation owin to the Heisenberg uncertainty principle. Quantization of the robot equations i arried out via the Hamiltonian formalism by adding quantum noise terms with indblad coefficients to the Schrodinger picture dynamics.

Quantization of any stochastic dynamical noisy system described by a classi-al stochastic differential equation is carried out by the method of Evans-Hudson lows wherein functions of the robot angular position and velocity are regarded s quantum observables and partial derivatives of such functions evaluated at the obot angular position and velocity are interpreted as algebra homomorphisms cting on linear transformations, ie, structure maps of the observables. In this vay, a classical stochastic differential equation becomes a quantum stochastic lifferential equation for the evolution of an algebra homomorphism. The prob-em of how to choose the structure maps of an Evans-Hudson flow so that the rocess has Gauss-Markov statistics in a given initial state is analyzed. Noisy lassical Hamiltonian dynamics can be cast in the form of a quantum Evans-Iudson flow in this way by applying the classical Ito formula for functions and hen replacing classical Brownian motion and Poisson processes by their non-ommutative quantum generalizations of the Hudson-Parthasarathy quantum tochastic calculus. The differential operators acting on function in the classi-al Ito theory can be generalized to structure maps acting on non-commuting bservables.

Another problem involves calculating the Green's function for the n-dimensional Ielmholtz equation using the theory of generalized functions. This has appli-ations to calculating the radiation field produced by a robot comprising of -dimensional links carrying current and moving in n dimensional space. This nables us to estimate the state of the classical robot as a function of time from he statistics of the measured radiation field.

Another problem addresses the isue of computing the conditional expecta-ion in quantum probability when the conditioning algebra is Abelian and this lgebra commutes with the observable whose conditional expectation we seek. This has application to estimating the state of the quantum robot from non-lemolition measurements. Conditional expectations can be defined only when oint probabilities exist and these exist only when the observables in question ll commute, otherwise Heisenberg's uncertainty principle will apply and pre-ent simultaneous measurablity of the conditioning algebra and the conditioned bservable.

Another issue is to define a quantum Markov process in discrete time using ompletely positive maps acting on an algebra of observables and an evolution or a family of algebra homomorphisms indexed by time with the evolution spec-fied by structure maps on the space of observables. Classical Markov processes hen arise as special commutative cases of this picture in which the observables orm a commutative algebra of functions and the family of homomorphisms cting on such a function coincides with evaluating the function at the current hase space of the classical robot. This is important in problems involving dis-retization of the dynamics of a robot followed by quantization resulting in the tate following a quantum Markov process in discrete time which is the quantum eneralization of the classical Markov process obtained by discretizing a classical tochastic differential equation followed by the robot with driving noise being any independent increment process.

Stochastic Lyapunov energy function of a robot based on using Ito's calculus for calcuating the average rate of energy increase has also been discussed in the context of the energy being defined as a tracking error energy. This enables one to design controllers for minimizing the rate of average error energy increase. In the absence of noise, the controllers can be designed so that the Lyapunov error energy decreases with time thereby ensuring asymptotic stability while in the presence of noise, there is always a positive contribution to the average error energy increase so we can only hope to decrease the rate of average error energy increase.

The quantum stochastic stability problem is also discussed. The observable satisfies a Heisenberg qsde which is in fact an Evans-Hudson flow that may be for example derived from the Hudson-Parthasarathy noisy Schrodinger evolution equation using the adjoint map of the unitary evolution acting on a system observable and the Lyapunov energy function is the average value of a quadratic from of the evolving observable in a coherent state of the system and bath. The aim here is to choose the Lindblad noise coefficient operators in system Hilbert space so as to reduce the rate of increase of the Lyapunov energy. The quadratic form may for example be defined as the square of the tracking error of the observable w.r.t a given observable trajectory.

The next problem is to derive using perturbation theory this shift in the modes of the electromagnetic field within a waveguide filled with an inhomogeneous medium and if the medium permittivity and permeablity fields have random components then we evaluate the statistical correlations of the perturbation of modal fields. This finds application in controlling the motion of a robot carrying current by causing the waveguide generated electromagnetic field to interact with the robot. If there are stochastic fluctuations in the guide fields caused by random media then correspondingly there will be stochastic fluctuations in the electromagnetically controlled robot motion and our guide must be designed to minimize these fluctuations. This problem can also be generalized to the quantum context where a quantum robot carrying a quantum current field interacts with the quantum electromagnetic field generated within a waveguide having a classical random medium and the aim is to control the quantum electromagnetic field within the guide as for example by changing the coherent state of the photons so that the classical plus quantum average of the robot position fluctuation around a given nonrandom classical trajectory in the state of the waveguide field and the robot current field is minimized.

Dirac equation based temperature estimation of blackbody radiation: Here, the statistical correlations of the electromagnetic four potential is such that the corresponding spectrum corresponds to that of homogeneous and isotropic blackbody radiation at a given temperature T. When a Dirac electron interacts with such an electromagnetic field, then the resulting wave function of the electron will have random fluctuations with satistics dependent upon the temperature of the black body radiation and hence by measuring the average of a quantum observable of the electron in such a pure state, we can estimate the temperature. The application of this to robotics is as follows. The robot consists of a single quantum rigid body and we write down it Hamiltonian as a function of the three Euler angles and the corresponding canonical angular momenta derived from the Lagrangian of a rigid body. However, we do this taking relativistic approximations into account while computing the kinetic energy. This relativistic Hamiltonian is approximately factorized as a product of terms that are linear in the canonical momenta and thus we may derive an approximate Dirac Hamiltonian of the rigid body. When this quantum rigid body interacts with external electromagnetic radiation owing to the charge on the body, then we can compute the Dirac evolution of the wave function of the rigid body as a function of the electromagnetic field and can hence by measuring the fluctuations in the averages of observables associated with this Dirac quantum robot hope to get a good estimate of the temperature of radiation. It is to be noted that as of now there is no literature on how to quantize the relativistic motion of a rigid body in a Lorentz covariant way. So this problem is just a start.

Stochastic optimal control of master and slave robots. The master robot controls the slave robot via an error feedback torque so that the slave robot responds to the torque applied by the master robot hand operator. Likewise the slave robot moves in an environment and sends via a feedback torque information about the environment in which it moves to the master torque. This setup is important for example in problems like surgery where the slave robot is small enough so that it can do surgery inside a living body while the master robot is large and can only control the slave robot based on the feedback it receives from the slave about the environment in which it moves. Apart from these feedback torques, there are also noise terms in the dynamics of the master and slave robot. The aim is to design using the stochastic Bellman-Hamilton-Jacobi optimal control equations the extra control torques depending upon the instantaneous states of the master and the slave so that the expected value of a cost function of the master and slave robot states is minimized.

We have a large number N of robots moving in space and rotating about their joints and also interacting with each other via some potentials, for example each robot could carry some charge and current which would generate an electromagnetic field that would interact with the charge and current in another robot. The initial configuration of these robots is random and using the Liouville equation in mechanics, we could write down an evolution equation for the joint probability density of the states of all the robots at a given time. We may even have noise in the dynamics like white Gaussian and Poisson noise in which case we could write down the Fokker-Planck equation for the joint probability density of all the robot states based on Ito's calculus. Then by averaging over the states of all but one of the robots, all but two of the robots etc and making appropriate approximations, we could arrive at Bolztmann like kinetic transport equations for the marginal probabilities. This example could be used to study the dynamics of a very large system of interacting robots taking internal configurations also into account. Just as Varadhan et.al have derived hydrodynamical scaling limits for interacting particle systems, we could do the same with the exception that here the density will be a function of not only the external position but also the internal coordinates specified by Euler angles.

One of the problems deals with analyzing how an incident electromagnetic field comprising a superposition of plane waves over all the directions at a given frequency interacts with an inhomogeneous medium and then the scattered wave field is obtained by using perturbation theory combined with the expression for the amplitude of the plane waves defined as a function of the direction in terms of the spherical harmonics. This has the following important application to robotics. When a robot carrying current interacts with such an electromagnetic field, its configuration changes and from the change in its configuration, it is easy to identify the coefficients in the expansion of the plane wave in terms of spherical harmonics. Basically, these coefficients can be looked upon as unknown parameters in a linear model to which the basics of linear statistical inference can be applied.

Equations of motion of a charged string interacting with an external electromagnetic field. The Lagrangian is set up taking the string kinetic energy, its potential energy of stretching, (the former is a quadratic form in the string velocity field and the latter is a quadratic form in the string position field. The electric field and magnetic field are assumed to be spatially constant so the electric field interaction with the string is linear in the string position field while the magnetic field interaction is linear in the string's angular momentum field. The latter follows because the magnetic vector potential corresponding to a spatially constant magnetic field is a linear function of position and the vector potential interacting with the string is a linear function of the string velocity. This has importance in robotics because a string can be used as a flexible robot to perform various kinds of jobs. Further, a charged string can be controlled by an electromagnetic field to stay in a certain configuration from where the charge on it will generate an electromagentic field which can be made to interact with the charge and current on a robot so as to control the robot's motion.

One of the important problems discussed in this book involves estimating the parameters of a robot that includes state dependent torque terms with unknown parameters and disturbance so that the robot follows a desired trajectory and simultaneously the disturbance estimation error with the estimate of the disturbance obtained by a standard instantaneous state dependent disturbance observer is minimized. This optimization problem is formulated as a least squares parameter estimation problem with the cost function being a sum over error squares upto the current time t. Such a formulation enables one to use the recursive least squares algorithm to obtain real time adaptive parameter estimates. Further, when there are random torque terms in the dynamics, we may linearize the robot differential equation about the noiseless trajectory thereby obtaining a linear stochastic differential equation for the robot state perturbation. By solving this using the state transition matrix method, we calculate the statistical correlations in the robot angular position and velocity processes and in the disturbance estimation error process which in turn can be used to determine the performance, ie, robustness of the the controlled robot to noise.

Variational principle for fields and their quantizations using the Feynman path integral. The classical Euler-Lagrange equations for fields are derived given the Lagrangian density with an interaction term between the fields and a current ource. Then, the Feynman path integral for such fields is formulated giving hereby the time ordered moments of the fields for a given external current ource. This is important in quantum robotics where the robots in motion provide a controllable classical current source which can be used to influence he interaction of elementary particles in an accelerator thereby enabling us to iscover new kinds of elementary particles as for example superpartners of the nown elementary particles.

Quantization of the equations of a classical robot with pd controllers. These quations cannot be derived from a Hamiltonian since the control forcing conains terms that are linear in the angular velocities/angular momenta. However, hese equations can be derived from the GKSL equation for open quantum sysems by an appropriate choice of the Lindblad operators. This is an important point from the philosophical standpoint of how to describe the motion of a quanum robot in a physically meaningful way. In the Hudson-Parthasarathy theory, here are quantum noise terms in the dynamics apart from the Hamiltonian part nd after tracing out over the bath state, the system state satisfies the GKSL quation. Owing to this fact, one can say that pd controllers of a quantum robot re actually applied to the bath. This fact has proven to be one of the triumphs f Belavkin's quantum filtering theory of using non-demolition measurements n the bath processes passed through the system in order to obtain an estimate f the system state dynamically in time.

General relavitistic quantum scattering: Studying standard problems in uantum scattering theory when general relativistic corrections are incorporated nto the Hamiltoinian of the projectile interacting with the scattering centre.

Design of robot controllers based on the recursive least squares algorithm ather than on the extended Kalman filter with a comparison of advantages and isadvantanges.

Motion of 3-D rigid bodies in general relativity. Techniques for writing down he Lagrangian of the rigid body in terms of rotation matrices for a given metric ield are developed.

Klein-Gordon field equation in the presence of a random electromagnetic ector potential field. How to evaluate the statistical moments of the Klein-Gordon field in terms of those of the electromagnetic field are discussed. This as important applications in large deviation theory (LDP)(where if the elecromagnetic field is a small random disturbance, then we can calculate the LDP ate function for the KG field to deviate from a stability zone and hence control he parameters of the KG system so as to minimize this probablity. In partile accelerators, this becomes important when we wish to control the rate of eaction between elementary particles resulting in modified rates for emission, bsorption and scattering of particles.

Study the interaction between a robot having 3-D links carrying current vith an external Klein Gordon field. The total Lagrangian of the KG field aking into account its interaction with the moving robot's current field and the agrangian of the robot taking into account the interaction of its current field vith the external KG field is set up. The resulting Euler-Lagrange equations letermine both the robot dynamics as well as the KG field dynamics in the presence of their mutual interactions. This setup can be used both to control the motion of a robot using an external KG field and also control the dynamics of the KG field in a particle accelerator by means of a robot.

Chapter 7

LDP in Spin Field Theory, Anharmonic Perturbations of Quantum Oscillators, Small Perturbations of Quantum Gibbs States

7.1 Large deviation problems in spin-field interaction theory

The Hamiltonian of an atom interacting with a magnetic field is

$$H_0 + e(\mathbf{B}(t,\mathbf{r}),\sigma)/2m$$

where $H_0 = \mathbf{P}^2/2m + V(\mathbf{r})$ is the Schrodinger Hamiltonian and the second term is the interaction Hamiltonian between the electron spin and a random external magnetic field $\mathbf{B}(t,\mathbf{r})$. Assuming the magnetic field to be of low amplitude, compute using N^{th} order perturbation theory the approximate transition probability of the atom between two of its stationary states and evaluate the LDP rate function for this transition probability in terms of the probability distribution of the magnetic field process in space-time. You can assume that the magnetic field is a Gaussian field.

Now, assume that the magnetic field is a quantum field described as a linear combination of creation and annihilation operators of the field and then calculate the approximate probability of atomic transitions with the field remaining in a fixed coherent state. When the field amplitude becomes small, evaluate the rate function of this transition probability in terms of the coherent state vector. Repeat this calculation when the field is in a superposition of coherent states.

Let $\sigma_1, ..., \sigma_N$ be independent Pauli spin matrix vectors acting in different components of a tensor product of the N copies of the Hilbert space $\mathbb{C}^2$ These spins are arranged in a circle with nearest neighbour interactions so that their interaction Hamiltonian is

$$H = \sum_{i=1}^{N} a(N,i)(\sigma_i, \sigma_{i+1}), \sigma_{N+1} = \sigma_1$$

Calculate the eigenvalues of H and the partition function

$$Z(\beta) = Tr(exp(-\beta H))$$

as a function of $a(N,i), i = 1,2,...,N$. Take an observable $X(N)$ in this tensor product Hilbert space and evaluate the probability distribution of $X(N)$ in the state

$$\rho(N) = exp(-\beta H)/Tr(exp(-\beta H))$$

Under what conditions on the real number sequence $a(N,i), i = 1,2,...,N, N = 1,2,...$ does this distribution display large deviation properties ?

7.2 Large deviation problems associated with a quantum gravitational field interacting with a non-Abelian gauge field

The quantum gravitational field can be described by a Hamiltonian that is a function of position and momentum fields with the position field being the spatial components of the metric tensor and the corresponding momentum fields being derived from the ADM action. Write this Hamiltonian after discretization as

$$H_{gN}(q_1, ..., q_N, p_1, ..., p_N)$$

As $N \to \infty$, we get a more and more accurate representation of the exact graviational Hamiltonian. Now suppose that this gravitational field interacts with a non-Abelian gauge field. The total Hamiltonian can be expressed as

$$H_N = H_{gN}(\mathbf{q}, \mathbf{p}) + H_{YN}(\mathbf{q}', \mathbf{p}') + H_{IN}(\mathbf{q}, \mathbf{q}', \mathbf{p}, \mathbf{p}')$$

Evaluate the mean value of any function of the non-Abelian gauge field in and more generally its probability distribution in the thermal state

$$\rho(N) = exp(-\beta H_N)/Tr(exp(-\beta H_N))$$

and obtain large deviation properties of this probability distribution function as $N \to \infty$.

.3 Large deviation problems in quantum harmonic oscillator problems with nonlinear terms

et $a_k, a_k^*, k = 1, 2, ..., p$ be annihilation and creation operators of independent armonic oscillators, ie,

$$[a_k, a_m^*] = \delta_{km}, [a_k, a_m] = [a_k^*, a_m^*] = 0$$

onsider the harmonic oscillator Hamiltonian with anharmonic perturbation rms:

$$H = H_0 + \epsilon.H_1$$

here

$$H_0 = \sum_{k=1}^{p} \omega(k) a_k^* a_k, H_1 = f(a_k, a_k^*, k = 1, 2, ..., p)$$

here f contains cubic and higher order terms. The Schrodinger equation for mixed state is

$$i\rho'(t) = [H_0, \rho(t)] + \epsilon.[H_1, \rho(t)]$$

xpand

$$\rho(t) = \pi^{-p} \int f(t, z, \bar{z}) |\phi(z)><\phi(z)| d^p z d^p \bar{z}$$

here $|\phi(z)>$ is a coherent state, ie,

$$a_k |\phi(z)>= z_k |\phi(z)>, a_k^* |\phi(z)>= \frac{\partial}{\partial z_k} |\phi(z)>$$

here

$$z = ((z_k))_{k=1}^p \in \mathbb{C}^p$$

erive from the Schrodinger equation, a partial differential equation satisfied the complex valued function $f(t, z, \bar{z})$. Of course, this will be parametrized the small perturbation parameter ϵ. So we can expand it as

$$f(t, z, \bar{z}, \epsilon) = f_0(t, z, \bar{z}) + \sum_{m=1}^{\infty} f_m(t, z, \bar{z}) \epsilon^m$$

ow calculate the probability of an event like say the system is in the number ate $|n_1, ..., n_N>$ or that the system is in a coherent state $|\phi(w)>$. These obabilities can be expressed respectively as

$$< n_1, ..., n_N |\rho(t)| n_1, ..., n_N >, < \phi(w) |\rho(t)| \phi(w) >$$

xpress these probabilities in the form

$$\sum_{m \geq 0} \epsilon^m < n_1, ..., n_N |\rho_m(t)| n_1, ..., n_N >,$$

$$\sum_{m\geq 0} \epsilon^m < e(w)|\rho_m(t)|e(w) >$$

where

$$\rho_m(t) = \pi^{-p} \int f_m(t,z,\bar{z})|\phi(z)>< \phi(z)|d^p z d^p \bar{z}, n \geq 0$$

From these expressions, try to obtain large deviation results as for example, the rate at which the probability of the event will converge to the corresponding probability of the same event in the absence of perturbation, ie, using only $\rho_0(t)$.

7.4 Formulation of an LDP for quantum stochastic processes

The large deviation principle in classical probability states that if $B(t)$ is Brownian motion then

$$t^{-1}.log\mathbb{E}exp(\int_0^t V(B(s))ds) \to \lambda_{max}(V)$$

where $\lambda_{max}(V)$ is the maximum eigenvalue of $(1/2)d^2/dx^2 + V(x)$. Further, introducing the rate function

$$I(\mu) = sup(\int V(x)d\mu(x) - \lambda_{max}(V))$$

it is known that $I(\mu)$ is the rate function of the empirical distribution of Brownian motion. This idea can be generalized to a large class of stochastic processes in place of Brownian motion and this class includes many Levy processes like the compound Poisson process and stable processes having characteristic function $exp(-t|\omega|^a), 0 < a < 2$. Can one obtain a corresponding quantum rate function for a quantum random process (non-commutative in time) ? Specifically, if $X(t), t \geq 0$ is a family of observables, and V a real valued function of a real variable, then what is the limit

$$lim_{t\to\infty} t^{-1} log(Tr(\rho.exp(\int_0^t V(X(s))ds))$$

for a given state ρ ? When $X(t) = j_t(X)$ is a quantum Markov process satisfying the Evans-Hudson flow equations in terms of quantum Brownian motion and quantum Poisson processes, then when does this limit exist ?

[55] Large deviations in the equilibrium distribution of a Markov chain. Let $X(t)$ be a continuous time Markov chain and let its infinitesimal generator matrix $((\lambda(i,j|\theta)))$ depend upon a parameter θ. A stationary distribution $\pi_i(\theta)$ for this chain satisfies

$$\sum_i \pi_i(\theta)\lambda(i,j|\theta) = 0 \forall j$$

Note that

$$\sum_j \lambda(i, j|\theta) = 0$$

Now when θ becomes a random parameter as in random environment problems, then can one derive statistical properties of the stationary distribution $\pi_i(\theta)$? and specifically, if θ_n is a sequence of random variables converging to θ_0, then at what rate does $\pi_i(\theta_n)$ converge to $\pi_i(\theta_0)$? where θ_0 is non-random ?

Chapter 8

LDP for Electromagnetic Control of Gravitational Waves, Randomly Perturbed Quantum Fields, Hartree-Fock Approximation, Renewal Processes in Quantum Mechanics

8.1 Gravitational wave propagating in a background curved space-time, LDP for reducing the wave fluctuations via electromagnetic control

Einstein's field equations in a control electromagnetic field are

$$G_{\mu\nu} = R_{\mu\nu} - (1/2)Rg_{\mu\nu} = K.S_{\mu\nu}$$

where

$$S_{\mu\nu} = (-1/4)F^{\alpha\beta}F_{\alpha\beta}g_{\mu\nu} + F_{\mu\alpha}F^{\alpha}_{\nu}$$

or equivalently, since the electromagnetic field energy-momentum tensor has zero trace,

$$R_{\mu\nu} = KS_{\mu\nu}$$

After linearizing this equation around a background metric $g^{(0)}_{\mu\nu}(x)$ and assuming that the electromagnetic source contributes only to the metric perturbations and not to the background gravitational field, we get

$$\delta R_{\mu\nu} = KS_{\mu\nu}$$

Now,

$$\delta R_{\mu\nu} = \delta\Gamma^{\alpha}_{\mu\alpha,\nu} - \delta\Gamma^{\alpha}_{\mu\nu,\alpha}$$

$$-\Gamma^{\alpha}_{\mu\nu}\delta\Gamma^{\beta}_{\alpha\beta} - \Gamma^{\beta}_{\alpha\beta}\delta.\Gamma^{\alpha}_{\mu\nu} + \Gamma^{\alpha}_{\mu\beta}\delta\Gamma^{\beta}_{\nu\alpha}$$

$$+\Gamma^{\beta}_{\nu\alpha}.\delta\Gamma^{\alpha}_{\mu\beta}$$

$$= (\delta\Gamma^{\alpha}_{\mu\alpha})_{:\nu} - (\delta\Gamma^{\alpha}_{\mu\nu})_{:\alpha}$$

where the covariant derivative is taken w.r.t the curved background metric. The resulting wave equation has the form

$$C_1(x\mu\nu\rho\sigma\alpha\beta)\delta g_{\rho\sigma,\alpha\beta}(x)+C_2(x\mu\nu\rho\sigma\alpha)\delta g_{\rho\sigma,\alpha}(x)+C_3(x\mu\nu\rho\sigma)\delta g_{\rho\sigma}(x) = K.S_{\mu\nu}(x$$

In this expression, the coefficient functions C_1, C_2, C_3 are expressible in terms of the metric tensor of the background curved space-time and its first and second order partial derivatives w.r.t the space-time indices. Now assume that the electromagnetic source $S_{\mu\nu}$ is a weak amplitude random field whose statistics is a Gaussian-Poisson mixture. Suppose $H(x)$ is a mixed Gaussian-Poisson field. We can express it as

$$H(x) = w(x) + \int_{\mathbb{R}^4} f(x,y)N(dy)$$

where $N(.)$ is a space-time Poisson field with intensity $dF(x)$, ie,

$$\mathbb{E}(N(E)) = F(E), P(N(E) = n) = exp(-F(E))F(E)^n/n!, n = 0, 1, ...$$

and $w(x)$ is a zero mean Gaussian field independent of $N(.)$ with correlation

$$\mathbb{E}(W(E)W(F)) = \mu(E \cap F), W(E) = \int_E w(x)dx$$

for $E, F \in \mathcal{B}(\mathbb{R}^4)$. So the above pde for the metric perturbation can be cast in the form

$$\mathcal{L}(x)(\psi(x)) = H(x)$$

where $L(x)$ is a second order linear partial differential operator acting on space-time functions. Let $Q(x,y)$ denote the inverse integral kernel of $L(x)$, ie,

$$\int Q(x,y)L(y)(f(y))dy = f(x) = L(x)(\int Q(x,y)f(y)dy)$$

Then, we can formally write

$$\psi(x) = \int Q(x,y)H(y)dy$$

The LDP rate function for this random field must be computed and then the the background gravitational field by altering its parameters so that the deviation probability of the metric perturbations defined by the function $\psi(x)$ from zero over a given space-time region by an amount more than a given threshold is minimized. For this computation, we require the moment generating functional of a Poisson field:

$$\mathbb{E}[exp(\int \phi(x)N(dx))] = exp(\int (exp(\phi(x)) - 1)dF(x))$$

When nonlinear terms are taken into account in the perturbed Einstein field equations, we obtain a pde of the form

$$A_1(x)(\nabla \otimes \nabla\psi(x)) + A_2(x)(\nabla\psi(x)) + A_3(x)\psi(x)$$

$$+B_1(x)((\nabla\otimes\nabla)\psi(x)\otimes\psi(x))+B_2(x)(\nabla\psi(x)\otimes\nabla\psi(x))+B_3(x)(\nabla\psi(x)\otimes\psi(x))+$$
$$B_4(x)(\psi(x)\otimes\psi(x)) = H(x)$$

and for this problem, we must determine the approximate rate function using perturbation theory. Here, $A_k(x), k = 1,2,3, B_k(x), k = 1,2,3,4$ are matrices of appropriate sizes.

8.2 The Lehmann representation of the propagator

$$< 0|\phi(x)\phi(y)|0 >= \int < 0|\phi(x)|p_1 >< p_1|\phi(y)|0 > d^3p_1$$
$$+\int < 0|\phi(x)|p_1p_2 >< p_1p_2|\phi(y)|0 > d^3p_1d^3p_2+...$$

the sum extending over one particle states, two particle states, and more generally, k particle states for $k = 1,2,....$ From such a representation, we can show that the propagator in the momentum domain can be expressed as a superposition of standard Klein-Gordon propagators over all masses:

$$G_\phi(p) - \int d\mu(m)/(p^2 - m^2), p = (p^\mu), p^2 = (p_0)^2 - (p^1)^2 - (p^2)^2 - (p^3)^2$$

Note that the propagator in the space-time domain is

$$G_\phi(x,y) = \theta(x^0 - y^0) < 0|\phi(x)\phi(y)|0 > +\theta(y^0 - x^0) < 0|\phi(y)\phi(x)|0 >$$

and we use

$$< 0|\phi(x)|p_1 >=< 0|\phi(x)a(p_1)^*|0 >= exp(-ip_1.x)u(p_1)$$

$$< 0|\phi(x)|p_1p_2 >=< 0|\phi(x)a(p_1)^*a(p_2)^*|0 >=$$

$$= \int <0|a(p)a(p_1)^*a(p_2)^*)|0> u(p)exp(-ip.x)d^3p$$

etc, where we use the mass shell expansions

$$\phi(x) = \int (a(p)u(p)exp(-ip.x)+a(p)^*u(p)^*exp(ip.x))d^3p, p^0 = \sqrt{P^2+m^2} = E(P$$

$$P = (p^1, p^2, p^3), u(p) = (2E(P))^{-1/2}$$

Note that

$$a(p)a(p_1)^*a(p_2)^*|0>= ([a(p), a(p_1)^*]a(p_2)^* + a(p_1)^*[a(p), a(p_2)^*])|0>$$
$$= (\delta^3(P-P_1)a(p_2)^* + \delta^3(P-P_2)a(p_1)^*)|0>$$
$$= \delta^3(P-P_1)|p_2> + \delta^3(P-P_2)|p_1>$$

and

$$<0|p>= \delta^3(P)$$

Problem: Evaluate

$$<0|\phi(x_1)...\phi(x_n)|0>, n \geq 1$$

Also evaluate

$$<p_1|\phi(x_1)...\phi(x_n)|p_2>=<0|a(p_1)\phi(x_1)...\phi(x_n)a(p_2)^*|0>$$

and more generally,

$$<p_1...p_r|\phi(x_1)...\phi(x_n)|p_1'...p_s'>=$$
$$<0|a(p_1)...a(p_r)\phi(x_1)...\phi(x_n)a(p_1')^*...a(p_s')^*|0>$$

and hence derive the Lehmann representation for the nth order propagator:

$$<0|T(\phi(x_1)...\phi(x_n))|0>$$

8.3 The LDP problem in this context

Consider now a KG field with a random Gaussian potential perturbation

$$[\nabla^2 - \partial_t^2 - m^2 - \epsilon V(x)]\phi(x) = 0$$

Evaluate the propagator for the corresponding quantum field assuming that the unperturbed field can be expanded in terms of standard KG creation and annihilation operators in the momentum domain. Evaluate this propagator as a power series in ϵ and derive the probability that this propagator will deviate from the standard free, ie, noiseless KG propagator by an amount greater than a given threshold.

8.4 Central limit theorem for renewal processes

Let $S_n = X_1 + ... + X_n$ be a renewal process, ie, the $X_i's$ are iid non-negative r.v's. Let $N(t) = max(n : S_n \leq t)$. Then

$$N(t) = \sum_n \chi_{S_n \leq t}$$

Let $\mathbb{E}(X_1) = m$. Then,

$$\{m - \epsilon < S_n/n \leq m + \epsilon\} = \{n(m - \epsilon) < S_n \leq n(m + \epsilon)\}$$
$$= \{N(n(m - \epsilon)) < n \leq N(n(m + \epsilon))\}$$

Equivalently,

$$\{N(t)/t \geq 1/m + \epsilon\} = \{N(t) \geq t(1/m + \epsilon)\}$$
$$= \{S_{[t(1/m+\epsilon)]} \leq t\}$$

and

$$\{N(t)/t < 1/m - \epsilon\} = \{N(t) < t(1/m - \epsilon)\}$$
$$= \{S_{[t(1/m-\epsilon)]} > t\}$$

From these equations and the strong law of large numbers $S_n/n \to ma.sP$, it follows that $N(t)/t \to 1/ma.s.P$.

Note that

$$N(t)/t = t^{-1} \sum_n \chi_{S_n \leq t}$$

Now we have for our renewal process, with $\sigma^2 = Var(X_1)$ that

$$(S_{N(t)} - mN(t))/\sigma\sqrt{N(t)}$$

converges in distribution as $t \to \infty$ to $N(0, 1)$. This because for large t, $N(t) \approx 1/mt$ and $\mathbb{E}(S_n) = nm, Var(S_n) = n\sigma^2$ and $(S_n - nm)/\sigma\sqrt{n}$ converges in distribution to $N(0, 1)$. Now we can rewrite this renewal theory version of the central limit theorem in the following way: $S_{N(t)}/N(t) \approx m$ and $N(t)/t \approx 1/m$ so $S_{N(t)} \approx N(t)m \approx t$ and hence

$$(mN(t) - t)/\sigma\sqrt{t/m} \to N(0, 1)$$

or equivalently,

$$(N(t) \quad t/m)/(\sigma m^{-3/2}\sqrt{t}) \to N(0, 1)$$

To prove this result rigorously, we must show that $(S_{N(t)} - t)/\sqrt{t} \to 0$, in distribution but this follows from the fact that $t - S_{N(t)}$ falls in the range $[0, X_{N(t)+1}]$ and hence the distribution of $S_{N(t)+1} - t$ is concentrated over support of F where F is the distribution of X_1. More precisely, $t - S_{N(t)} \geq 0$ and for $x > 0$,

$$P(|t - S_{N(t)}| > x) = P(t - S_{N(t)} > x) \leq P(X_1 > x) = 1 - F(x)$$

which implies that

$$P(|t - S_{N(t)})/\sqrt{t} > x) \leq 1 - F(x\sqrt{t}) \to 0, t \to \infty$$

thereby completing the proof of the central limit theorem for renewal processes.

8.5 Applications of renewal process theory in quantum field theory

The photons hit a conducting plate at times $S_n, n = 1, 2, ...$ forming a renewal process. The number of photons that hit the plate in time $[0, t]$ is $N(t)$ which is the renewal counting process. Each time that a photon hits the plate, it generates a current $h(t - \tau)$ where τ is the hitting time of that photon. The current generated in the plate at time t is therefore given by

$$I(t) = \sum_k h(t - S_k) = \int h(t - \tau)dN(\tau)$$

Now this current interacts with a quantum electromagnetic field or more generally with a non-Abelian gauge field as follows: First this current feeds into an antenna thereby generating a current density

$$J^{\mu a}(t, r) = \int F^{\mu a}(t - \tau, r)I(\tau)d\tau$$

which in the temporal Fourier transform domain reads

$$\hat{J}^{\mu a}(\omega, r) = \hat{F}^{\mu a}(\omega, r)\hat{I}(\omega)$$

ie there is a transfer function for each spatial point in the antenna which takes as input the current at its feed and generates a current density at that spatial point. The Lagrangian density for the non-Abelian matter and gauge fields when they interact with this antenna current is given by

$$L = (-1/4)F^a_{\mu\nu}F^{\mu\nu a} + \psi^*\gamma^0(\gamma^\mu(i\partial_\mu + eA^a_\mu\tau_a) - m)\psi - J^{\mu a}A^a_\mu$$

where τ_a are the Hermitian generators of the gauge group. The equations of motion for the matter and gauge field are then

$$D_\nu F^{\mu\nu a} = J^{\mu a} - e\psi^*(\gamma^0\gamma^\mu \otimes \tau^a)\psi$$

$$[\gamma^\mu(i\partial_\mu + eA^a_\mu\tau_a) - m]\psi = 0$$

The latter matter field equation can also be expressed as

$$(\gamma^\mu i\partial_\mu - m)\psi = -eA^a_\mu(\gamma^\mu \otimes \tau^a)\psi$$

Now we can also conceive of a random gauge field source which will be a continuous function of time and space and hence can be expressed as a stochastic integral w.r.t Brownian motion

$$C^a_\mu(t, r) = \int \chi^a_\mu(r, t - \tau)dB(\tau)$$

Then this classical control gauge field must be added to the quantum gauge field A^a_μ so that the equations of motion of the matter field gets modified by the presence of such a source term

$$(\gamma^\mu i\partial_\mu - m)\psi = -e(A^a_\mu + C^a_\mu)(\gamma^\mu \otimes \tau^a)\psi$$

low suppose we modify the problem by replacing the random classical current nd field sources by their quantum noisy versions so that if $\Lambda(t)$ denotes the hoton counting process (ie, the conservation process in the language of Hudon and Parthasarathy) and $A(t), A(t)^*$ are annihilation and creation processes gain in the sense of Hudson and Parthasarathy, then

$$J^a_\mu(t,r) = \int G^a_\mu(t-\tau,r)d\Lambda(\tau)$$

$$C^a_\mu(t,r) = \int [\chi^a_\mu(t-\tau,r)dA(\tau) + \bar{\chi}^a_\mu(t-\tau,r)dA(\tau)^*]$$

lore generally, we can consider the Hudson-Parthasarathy quantum noise proesses $\Lambda^b_a(t)$ that satisfy the quantum Ito formula

$$d\Lambda^b_a(t).d\Lambda^d_c(t) = \epsilon^b_c d\Lambda^d_a(t)$$

nd define the quantum noisy current and gauge potential sources

$$J^a_\mu(t) = \int F_\mu(a,b,c,t-\tau,r)d\Lambda^c_b(\tau),$$

$$C^a_\mu(t) = \int \chi_\mu(a,b,c,t-\tau,r)d\Lambda^c_b(\tau)$$

he zeroth order solution to the quantum fields $\psi(t,r), A^a_\mu(t,r)$ is given by (ie 1 the absence of sources, interactions and nonlinearities)

$$_0(t,r) = \int [u(P,\sigma)a(P,\sigma)exp(-i(E(P)t-P.r))+v(P,\sigma)b(P,\sigma)^*exp(i(E(P)t-P.r))]d^3P,$$

$$^a_{0\mu}(t,r) = \int [(2|K|)^{-1/2}e^a_\mu(K,s)c(K,s)exp(-i(|K|t-K.r))+(2|K|)^{-1/2}\bar{e}^a_\mu(K,s)c(K,s)^*exp(i(|K|t-K.r))]d^3r$$

pplication of first order perturbation theory gives the following quantum noisy lus nonlinearity corrections to these quantum fields:

$$D_\nu F^{\mu\nu a} = \partial_\nu F^{\mu\nu a} + eC(abc)A^b_\nu F^{\mu\nu c}$$

here

$$F^{\mu\nu a} = \partial^\mu A^{\nu a} - \partial^\nu A^{\mu a} + eC(abc)A^{\mu b}A^{\nu c}$$

$$-\Box A^{\mu a}_1 + \partial^\mu(\partial_\nu A^{\nu a}_0) + eC(abc)\partial_\nu(A^{\mu b}_0 A^{\nu c}_0) + eC(abc)A^b_{0\nu}(\partial^\mu A^{\nu c}_0 - \partial^\nu A^{\mu c}_0$$

$$+eC(cdf)A^{\mu d}_0 A^{\nu f}_0) =$$

$$J^{\mu a} - e\psi^*_0(\gamma^0\gamma^\mu \otimes \tau^a)\psi_0$$

$$[i\gamma^\mu\partial_\mu - m]\psi_1 = -e(C^a_\mu + A^a_{0\mu})(\gamma^\mu \otimes \tau^a)\psi_0$$

olving these first order perturbed equations using the Green's functions, ie, he electron and linearized gauge boson propagators gives us an expression or the perturbations $A^{\mu a}_1, \psi_1$ as term that are linear in the source fields and onlinear in the zeroth order fields, or equivalently in the Fermion and boon creation and annihilation operators. Note that the solution for $A^{\mu a}_1$ will contain linear, quadratic and cubic combinations of the boson creation and annihilation operators $c(K,s), c(K,s)^*$ and also terms that are linear in the source current or equivalently in the quantum stochastic processes and also terms that are quadratic in the Fermion creation and annihilation operators $a(P,\sigma)a(P,\sigma)^*, b(P,\sigma), b(P,\sigma)^*$. On the other hand, the solution for ψ_1 will contain terms that are linear in the source gauge field or equivalently in the quantum stochastic processes bilinearly coupled to the Fermion creation and annihilation operators plus terms that are linear in the gauge boson creation and annihilation operators bilinearly coupled to the Fermion creation and annihilation operators. From these expressions, we can calculate the matrix elements of the perturbed quantum fields relative in states that are tensor products of the gauge boson coherent states, the fermion coherent states and the coherent states of the quantum stochastic bath. From these matrix elements, we can calculate in particular transition probabilities between two states of the system caused by the interaction Hamiltonian between the quantum gauge fields and the classical/quantum noisy current source and between the quantum gauge fields and the quantum non-Abelian matter field Dirac current and between the classical/quantum noisy gauge fields and the quantum non-Abelian matter field Dirac currents and by assuming these to be small, we can evaluate in principle the rate function for these transition probabilities.

Remark: If higher order perturbation theory is considered, then the quantum noisy sources will generate contributions to the quantum fields of the form of chaos integrals:

$$F(t) = \int_{0<t_1<...<t_n<t} \chi(\mathbf{a},\mathbf{b},t,t_1,...,t_n) d\Lambda_{b_1}^{a_1}(t_1)...d\Lambda_{b_n}^{a_n}(t_n)dt_1...dt_n$$

with the sum being over the indices $\mathbf{a} = (a_1, ..., a_n)$ and $\mathbf{b} = (b_1, ..., b_n)$. Evaluating the matrix element of this in the coherent states $|\phi(u)>, |\phi(v)>$ gives us noting that

$$d\Lambda_b^a(t) = dA_b(t)^* dA_a(t)/dt, A_0(t) = A_0(t)^* = t$$

that

$$< \phi(v)|F(t)|\phi(u) >= \int_{0<t_1<...<t_n<t} \chi(\mathbf{a},\mathbf{b},t,t_1,...,t_n) u_{a_1}(t_1)$$
$$...u_{a_n}(t_n)\bar{u}_{b_1}(t_1)......\bar{u}_{b_n}(t_n)dt_1...dt_n$$

where

$$u_0(t) = t$$

8.6 Large deviation principle in the Hartree-Fock method for approximately solving many electron problems

Let $H_a, a = 1,2,...,N$ be the Hamiltonians of each of the N electrons interacting with the nucleus and let $V_{ab}, a < b$ be their pariwise interaction potential energies. The total Hamiltonian is then

$$H = \sum_a H_a + \sum_{a<b} V_{ab}$$

To apply the variational principle to approximately calculate the stationary state eigenfunctions, we assume a trial joint wave function of the form

$$|\psi>= \sum_\sigma sgn(\sigma)\phi_{\sigma 1} \otimes ...\phi_{\sigma N}$$

and choose the $\phi_k' s$ so that

$$<\psi|H|\psi>$$

is minimized subject to the normalization condition $<\phi_k, \phi_k >= 1$ for all k. This Taking into account these constraints using Lagrange multipliers gives us the functional to be minimized as

$$<\psi|H|\psi> - \sum_{k=1}^N E_k(<\phi_k, \phi_k> -1)$$

and carrying out the variational process keeping in mind that

$$<\phi_{\sigma 1} \otimes ... \otimes \phi_{\sigma N}|H_k|\phi_{\rho 1} \otimes ... \otimes \phi_{\rho N}>$$

$$=<\phi_{\sigma k}|H_k|\phi_{\rho k}> \Pi_{j\neq k}\delta[\sigma j - \rho j]$$

and for $a < b$,

$$<\phi_{\sigma 1} \otimes ... \otimes \phi_{\sigma N}|V_{ab}|\phi_{\rho 1} \otimes ... \otimes \phi_{\rho N}>$$

$$=<\phi_{\sigma a} \otimes \phi_{\sigma b}|V_{ab}|\phi_{\rho a} \otimes \phi_{\rho b}> \Pi_{j\neq a,b}\delta[\sigma j - \rho j]$$

The result of carrying out this minimization is a sequence of nonlinear eigen equations for the individual wave functions $\phi_k, k = 1, 2, ..., N$. Now when a small random time independent external field is applied to this system of electrons and the nucleus assuming that this field is small, we can calculate the change in the stationary state wave function using time independent perturbation theory treating the unperturbed solution to be the Hartree-Fock approximate solutions. From these perturbed solutions, we can in principle calculate the approximate change in the average value of an observable defined on the tensor product Hilbert space and then evaluate the approximate probability that this average observable will fall in a certain domain using the LDP.

8.7 LDP in fuzzy neural networks

Let $\phi(x)$ be the fuzzification function. This is a map from $\mathbb{R}^p$ to $\mathbb{R}$ which is a standard p-variate Gaussian function. Now $f(x, m, S) = \phi((x-m)^T S^{-1}(x-m))$ is a Gaussian function with mean vector m and covariance matrix S. In the first layer of the network, we apply the input $x = (x(i))$ and the output of this layer which is the input to the second layer is

$$u(i) = f(x, m_i, S_i), i = 1, 2, ..., p$$

In other words, $u(i)$ is significant only if x is close to m_i within a spread defined by the covariance S_i. Here, $m_i, i = 1, 2, ..., p$ are "mean vectors" and $S_i, i = 1, 2, ..., p$ are positive definite matrices. The weight update of the next layer is given by

$$\phi(i, t+1) = (1 - \lambda(i, t))u(i, t) + \lambda(i, t)\phi(i, t)$$

Here, $\lambda(i, t) \in [0, 1]$ is another weight. This equation states that if $u(i, t)$ is large, which will happen if x is close to m_i within a mean square spread of S_i, then the first term on the rhs will be large causing the ϕ to increase provided that $\lambda(i, t)$ is small. If $\lambda(i, t)$ is large and $u(i, t)$ is small, the second term will dominate and will cause ϕ to evolve naturally at a decaying exponential rate. This is because if $u(i, t)$ is small, then it means that x is not close to m_i and hence does not correspond to the naturally required input and hence ϕ should not increase rapidly. The final choice of how λ evolves is dictated by the change in the output error energy.

Large deviations: Assume a neural firing model of the form

$$W(i, t+1) = \lambda(W_i(0) - W(i, t)) - \sum_s a(s)u(t-s)\partial(F(x(s), W(s)) - d(s))^2/\partial W(i, s$$

where $F(x(t), W(t))$ is the neural network output at time s and $d(s)$ is the desired output at time s. In vector notation, this evolution equation can be expressed as

$$\mathbf{W}(t+1) = \lambda(\mathbf{W}_0 - \mathbf{W}(t)) - \sum_{s<t} a(s)u(t-s)\nabla_{\mathbf{W}(s)}(F(\mathbf{x}(s) - \mathbf{W}(s)) - d(s))^2$$

$$= \lambda(\mathbf{W}_0 - \mathbf{W}(t)) - \sum_{s>0} a(t-s)u(s)\nabla_{\mathbf{W}(t-s)}(F(\mathbf{x}(t-s) - \mathbf{W}(t-s)) - d(t-s))^2$$

where now $u(t)$ is a decaying version of the unit step function, ie, $u(t)$ can be for example $exp(-at)\theta(t)$. As an example of such a firing application, consider a quantum neural network used to estimate a probability density function $p(t, x)$ evolving in time using the modulus square of the wave function satisfying Schrodinger's wave equation. The wave equation is

$$i\partial_t\psi(t, x) = -\partial_x^2\psi(t, x)/2m + V(t, x)\psi(t, x)$$

where

$$V(t, x) = W(t, x)(p(t, x) - |\psi(t, x)|^2)$$

with the weight $W(t, x)$ adapted as follows

$$\partial_t W(t, x) = \beta(W_0 - W(t, x)) - \mu(p(t, x) - |\psi(t, x)|^2)$$

he speed of this algorithm can be improved by including firing terms in the veight update based on errors between the desired pdf and the pdf generated y Schrodinger's equation. Here, Schrodinger's equation is the neural network. t is a recurrent neural network since the output wave function generated by it lepends on the past values of the wave function as well as the the past weights vith unit memory provided that we approximate the wave function derivative y a finite difference:

$$\delta^{-1}(\psi(t+1,x)-\psi(t,x)) = (-1/2m\Delta^2)(\psi(t,x+1)-2\psi(t,x)+\psi(t,x-1))+V(t,x)\psi(t,x)$$

uch a discretized equation does not guarantee unitary evolution and hence to nake the evolution unitary, we use the following approximation:

$$\psi(t+1) = U(t)\psi(t)$$

vhere $\psi(t) = ((\psi(t,x))_x$ is a column vector and $U(t)$ is a unitary matrix which an be constructed as

$$U(t) = exp(-i\delta H)$$

vith H a tridiagonal Hermitian matrix that is a spatially discretized from of he operator

$$(-1/2m)\partial_x^2 + V(t,x), V(t,x) = W(t,x)(p(t,x) - |\psi(t,x)|^2)$$

pecifically, the multiplication operator $W(t,x)(p(t,x) - |\psi(t,x)|^2)$ is replaced y a diagonal matrix while the partial differential operator ∂_x^2 is replaced by a ridiagonal matrix. A more advanced version of this qnn is obtained by considring the potential to be of the form

$$V(t,x) = \int W(t,x,y)(p(t,y) - |\psi(t,y)|^2)dy$$

$$i\partial_t\psi(t,x) = (-1/2m)\partial_x^2\psi(t,x) + V(t,x)\psi(t,x)$$

r more generally, by defining a mean square error at time t as

$$E(t) = \int a(t,x)(p(t,x) - |\psi(t,x)|^2)^2dx$$

nd defining the weight update as

$$W(t+1,x,y) = W(t,x,y) - \mu(\partial/\partial W(t,x,y))E(t+1)$$

vhere

$$\psi(t+1) = U(t)\psi(t), U(t) = exp(-i\delta H(t))$$

vith $H(t)$ being a tridiagonal Hermitian matrix defined as a spatially discretized ersion of

$$(-1/2m)\partial_x^2 + V(t,x) = (-1/2m)\partial_x^2 + \sum_y W(t,x,y)(p(t,y) - |\psi(t,y)|^2)$$

8.8 Large deviation analysis of this qnn

Assume that there is a small random perturbation $\delta p(t,x)$ to $p(t,x)$. The aim is then to determine the large deviation properties of the weights and hence of the approximating wave function $\psi(t,x)$ that solves the Schrodinger equation The solution to the problem can be approximately obtained by looking at the perturbed form of Schrodinger's equation

$$i\partial_t \delta\psi(t,x) = (-1/2m)\partial_x^2 \delta\psi(t,x) + \delta V(t,x)\psi(t,x) + V(t,x)\delta\psi(t,x)$$

where

$$\delta V(t,x) = W(t,x)\delta p(t,x) + \delta W(t,x)(p(t,x) - |\psi(t,x)|^2)$$

This gives us a stochastic partial differential equation for $\delta\psi(t,x)$.

8.9 LDP problems in quantum field theory related to corrections to the electron, photon and non-Abelian gauge boson propagators

The effect of non-Abelian gauge fields and the gravitational field on the electron's mass.

Chapter 9

LDP in Electromagnetic Scattering and String Theory, Control of Dynamical Systems Using LDP

9.1 A summary of a list of LDP applications in physics and engineering

1] If the finite dimensional distributions of a stochastic process vary by small andom amounts, then what is the rate function associated with the variation)f the probability measure of the process on path space ?

[2] If a measure-preserving transformation on a measure space changes by a ımall random amount, then what is the rate function associated with the time ıverage of an observable, ie, of a random variable defined on the measure space ' Specifically, let T denote the measure preserving transformation and let $f(\omega)$)e the observable assumed to be integrable. Its time average is

$$<f>(\omega) = lim_{n\to\infty} n^{-1}\sum_{k=0}^{n-1} f(T^k\omega)$$

ınd we know from Birkhoff's individual ergodic theorem that this equals $\mathbb{E}(f|I)(\omega)$ ı.s.P where I is the invariant σ-algebra, ie, $I = \{E \in \mathcal{F} : T^{-1}(E) = E\}$. The question is that when T changes to $S = T + \delta T$ which is also measure)reserving where we are assuming that the probability space is the sequence ıpace $(\mathbb{R}^{\mathbb{Z}_+}, \mathcal{B}(\mathbb{R}^{\mathbb{Z}_+}), P)$ associated with a stationary stochastic process, then ıow much does the time average change by ? Note that T is the unit shift ransformation but S need not be the shift transformation.

[3] If an electromagnetic field is confined within a cavity with the cavity having a permittivity field $\epsilon(\omega, r)$ and a permeability field $\mu(\omega, r)$, then calculate the energy of the field within the cavity using first order perturbation theory when

$$\epsilon(\omega, r) = \epsilon_0(1 + \delta.\chi(\omega, r)),$$

$$\mu(\omega, r) = \mu_0(1 + \delta.\chi_m(\omega, r))$$

The field energy is to be calculated upto $O(\delta)$ by solving the Maxwell equations in the frequency domain:

$$div(\epsilon E) = 0, div(\mu H) = 0, curlE = -j\omega\mu H, curlH = J + j\omega\epsilon E$$

perturbatively w.r.t the perturbation parameter δ and applying the energy conservation principle

$$\partial_t \int_V u(t, r)dV = -\int_S S(t, r).ndA - \int_V J(t, r).E(t, r)dV$$

in the time domain.

[4] Taking into account general relativistic corrections due to the background curved metric, set up the Maxwell equations as

$$F_{\mu\nu} = A_{\nu,\mu} - A_{\mu,\nu}, F^{\mu\nu} = g^{\mu\alpha}g^{\nu\beta}F_{\alpha\beta},$$

$$F^{\mu\nu}_{:\nu} = -\mu_0 J^\mu$$

The rate of power dissipation in the electromagnetic field per unit volume is given by $(\rho E + J \times B).v = J.E$ in special relativity which in general relativity is the zeroth component of the four vector $Q^\mu = F^{\mu\nu}J_\nu$. The energy density of the electromagnetic field and the energy flow per unit time per unit area, the momentum density and the momentum flow per unit time per unit area form components of the energy-momentum tensor of the electromagnetic field

$$S^{\mu\nu} = (-1/4)F_{\alpha\beta}F^{\alpha\beta}g^{\mu\nu} + F^{\mu\alpha}F^{\nu}_{\alpha}$$

and we have the energy-momentum conservation equation

$$S^{\mu\nu}_{:\nu} + Q^\mu = 0$$

The problem is to compute the deviation in the energy-momentum tensor of the electromagnetic field from that of special relativity caused by small perturbations to the flat space-time metric and then assuming that these metric perturbations are small and random, evaluate the large deviation rate function of the perturbed energy-momentum tensor.

9.2 LDP problems related to scattering of electromagnetic waves by a perfectly conducting cylinder

The cylinder surface is $0 \leq L, \rho = R$. A plane electromagnetic wave

$$\mathbf{E}_i = \mathbf{E}_{i0} exp(-ik.r), \mathbf{H}_i = \mathbf{H}_{i0} exp(-ik.r)$$

at frequency $\omega = |k|c$ is incident upon this cylinder. Let

$$\mathbf{J}_s = J_{s\phi}(\phi, z)\hat{\phi} + J_{sz}(\phi, z)\hat{z}$$

be the induced surface current density on the cylindrical surface. Maxwell's equations in a weakly curved space-time are

$$curlE = -B_{,t}, divB = 0,$$

$$(F^{\mu\nu}\sqrt{-g})_{,\nu} = J^\mu \sqrt{-g}$$

or writing

$$g_{\mu\nu} = \eta_{\mu\nu} + h_{\mu\nu}, g = -1 + h, \sqrt{-g} = 1 - h/2, h = \eta_{\mu\nu} h_{\mu\nu}$$

we get

$$F^{\mu\nu}_{,\nu} = (1/2)(hF^{\mu\nu}))_{,\nu} + J^\mu - (h/2)J^\mu$$

or using first order perturbation theory, with $F = F^0 + F^1$,

$$F^{0\mu\nu}_{,\nu} = J^\mu, F^{1\mu\nu}_{,\nu} = (hF^{0\mu\nu})_{,\nu} - (h/2)J^\mu$$

Writing

$$F^{0\mu\nu} + F^{1\mu\nu} = F^{\mu\nu} = g^{\mu\alpha} g^{\nu\beta} F_{\alpha\beta} =$$
$$(\eta_{\mu\alpha} - h^{\mu\alpha})(\eta_{\nu\beta} - h^{\nu\beta})(F^0_{\alpha\beta} + F^1_{\alpha\beta})$$
$$(\eta_{\mu\alpha}\eta_{\nu\beta} F^0_{\alpha\beta})$$
$$-(\eta_{\mu\alpha} h^{\nu\beta} + \eta_{\nu\beta} h^{\mu\alpha})F^0_{\alpha\beta}$$
$$+\eta_{\mu\alpha}\eta_{\nu\beta} F^1_{\alpha\beta}$$

we get

$$F^{0\mu\nu} = \eta_{\mu\alpha}\eta_{\nu\beta} F^0_{\alpha\beta}$$

$$F^{1\mu\nu} = -(\eta_{\mu\alpha} h^{\nu\beta} + \eta_{\nu\beta} h^{\mu\alpha})F^0_{\alpha\beta}$$
$$+\eta_{\mu\alpha}\eta_{\nu\beta} F^1_{\alpha\beta}$$

We can and shall indeed always choose our coordinate system so that $h_{0\mu} = 0, \mu = 0, 1, 2, 3$. Identifying the components of $F_{\mu\nu}$ with the true electric and magnetic field because it satisfies the usual homogeneous Maxwell equations

$$F_{\mu\nu,\alpha} + F_{\nu\alpha,\mu} + F_{\alpha\mu,\nu} = 0$$

or equivalently

$$divB = 0, curlE + \partial_t B = 0$$

or equivalently, in terms of potentials

$$F_{\mu\nu} = A_{\nu,\mu} - A_{\mu,\nu}$$

we get

$$F_{0r} = E_r, r = 1,2,3, F_{12} = -B_3, F_{23} = -B_1, F_{31} = -B_2$$

or equivalently,

$$B_r = -\epsilon(rsm)F_{sm}, r = 1,2,3$$

$$F^{00r} = -E_r^0, r = 1,2,3, F^{0rs} = F_{rs}^0 = -\epsilon(rsm)B_m^0, r, 1,2,3$$

$$F^{10r} = -F_{0r}^1 - h^{rs}F_{0s}^0 = E_r^1 + h^{rs}E_s^0 = E_r^1 + h_{rs}E_s^0$$

$$F^{1rs} = -F_{rs}^1 + h^{sm}F_{rm}^0 + h^{rm}F_{ms}^0$$

or equivalently, in terms of components,

$$F^{112} = B_3^1 + h^{22}F_{12}^0 + h^{23}F_{13}^0 + h^{11}F_{12}^0 + h^{13}F_{32}^0$$

$$= B_3^1 - h^{22}B_3^0 + h^{23}B_2^0 - h^{11}B_3^0 - h^{13}B_1^0$$

$$F^{123} = B_1^1 + h^{31}F_{21}^0 + h^{33}F_{23}^0 + h^{22}F_{23}^0 + h^{21}F_{13}^0$$

$$= B_1^1 + h^{31}B_3^0 - h^{33}B_1^0 - h^{22}B_1^0 + h^{21}B_2^0$$

Likewise for F^{131}:

$$F^{131} =$$

$$F^{131} = -F_{31}^1 + h^{1m}F_{3m}^0 + h^{3m}F_{m1}^0$$

$$= B_2^1 + h^{11}F_{31}^0 + h^{12}F_{32}^0 + h^{32}F_{21}^0 + h^{33}F_{31}^0$$

$$B_2^1 - h^{11}B_2^0 + h^{12}B_1^0 + h^{32}B_3^0 - h^{33}B_2^0$$

When the metric perturbations $h_{rs}(x)$ are weak amplitude zero mean Gaussian random fields, the problem is to calculate the rate function for the perturbed electric and magnetic fields both incident and scattered and also the rate function for the induced surface current density on the conducting cylinder.

3 String theory and large deviations

nsider a quantum string field

$$X^\mu(\tau,\sigma) = x^\mu + p^\mu\tau + i\sum_{n\neq 0} a^\mu(n)exp(in(\tau-\sigma))/n$$

$$+i\sum_{n\neq 0} b^\mu(n)exp(in(\tau+\sigma))/n$$

ere

$$a^\mu(-n) = a^\mu(n)^*, b^\mu(-n) = b^\mu(n)^*, n \neq 0,$$

$$[a^\mu(n), a^\nu(m)] = \eta^{\mu\nu} n\delta[n+m],$$

$$[b^\mu(n), b^\nu(m)] = \eta^{\mu\nu} n.\delta[n+m]$$

$$[a^\mu(n), b^\nu(m)] = 0$$

e center of this quantum string at time $\tau = 0$ is x^μ and the string perturbation time $\tau = 0$ is

$$\delta x^\mu(\sigma) = \sum_{n\neq 0} a^\mu(n)exp(-in\sigma)/n + \sum_{n\neq 0} b^\mu(n)exp(in\sigma)/n$$

e string propagator is easily computed to be

$$< 0|T(\delta x^\mu(\tau,\sigma)\delta x^\nu(\tau',\sigma'))|0 >=$$

$$\eta^{\mu\nu}.ln(\sqrt{(\tau-\tau')^2-(\sigma-\sigma')^2}) = \Delta^{\mu\nu}(\tau,\sigma|\tau',\sigma')$$

. Consider a field $\psi(x)$ as a function of the space-time coordinate $x = (x^\mu)$. ppose that it has an action functional

$$S[\psi] = \int L(\psi(x), \partial_\mu\psi(x))d^4x$$

wish to calculate the correction to this action functional due to the replacement of space-time points x by strings. To do so, we defined the string theoretic eraged action by

$$S_1[\psi] = (2\pi)^{-1} < \int_{0\leq\sigma<2\pi} L(\psi(x+\delta x(\sigma)), \partial_\mu\psi(x+\delta x(\sigma)))d^4x d\sigma >$$

ere $< . >$ denotes average value in say a coherent state of the harmonic osators comprising the string expansion. We make a linearized approximation:

$$L(\psi(x+\delta x(\sigma)), \psi_{,\mu}(x+\delta x(\sigma)) =$$

$$L(\psi(x) + \psi_{,\mu}(x)\delta x^\mu(\sigma), \psi_{,\mu}(x) + \psi_{,\mu\nu}(x)\delta x^\nu(\sigma))$$

$$L(\psi(x),\psi_{,\mu}(x)) + L_{,1}(\psi(x),\psi_{,\rho}(x))\psi_{,\mu}(x)\delta x^\mu(\sigma) + L_{,2\mu}(\psi(x),\psi_{,\rho}(x))\psi_{,\mu\nu}(x)\delta x^\nu(\sigma)$$

where we use the notation

$$L_{,2\mu}(\psi,\psi_{,\rho}) = \partial L(\psi,\psi_{,\rho})/\partial\psi_{,\mu}$$

We can now using this calculate the first order string theoretic correction to the action functional by using the coherent state formula

$$<\phi(u)|a^{\mu}(n)/\sqrt{n}|\phi(u)>= u^{\mu}(n), \bar{u}^{\mu}(n) = u^{\mu}(-n)$$

Now we describe a quantum mechanical path integral/averaging method for computing the string theoretic corrections to the effective action of a field. Let $\psi(x)$ be a field and $L(\psi(x), \psi_{,\mu}(x))$ the Lagrangian density of the field. The averaged string theoretic action for the field $S_1[\psi]$ is given by

$$exp(iS_1[\psi]) = \int exp(i(2\pi)^{-1}\int L(\psi(x+\delta x(\sigma)), \partial_\mu\psi(x+\delta x(\sigma))d^D x\sigma)\Pi_{0\leq\sigma<2\pi}D\delta x(\sigma))$$

$$= \int exp(i(2\pi)^{-1}\int L(\psi(x-i\sum_{n\neq 0}(\alpha(n)/n)exp(-in\sigma), \partial_\mu\psi(x-i\sum_{n\neq 0}(\alpha(n)/n)exp(-in$$

$$\times d^D x d\sigma)\Pi_{n>0}d\alpha(n)d\bar{\alpha}(n)$$

where now the $\alpha(n)'s$ are complex numbers with $\alpha(-n) = \bar{\alpha}(n)$.

The LDP problem: Assume that there are random parameters in the Lagrangian L of the point field. Then, determine the rate function of the string effect corrected action $S_1[\psi]$ and hence deduce what effect will these random parameters have on the quantum equations of motion based on the action $S_1[\psi]$.

The E_8 theory: Consider the group $SO(16)$. Its Lie algebra has dimension 15.16/2 = 120. which means that $SO(16)$ is generated by 120 linearly independent generators $\{X_1, ..., X_{120}\}$. Now consider the spin representation of $SO(16)$. This is obtained as follows. Let $V = \mathbb{C}^8$ and consider the vector space $\Lambda V = \bigoplus_{k=0}^{8} \Lambda^k V$. We clearly have $dim(\Lambda V) = 2^8$. We can define a basis of 8 Fermionic creation operators and their adjoints, namely a set of 8 linearly independent Fermion annihilation operators. Denote the former by $a_k^*, k = 1, 2, ...8$ and the latter by $a_k, k = 1, 2, ..., 8$. They satisfy the canonical anticommutation relations

$$\{a_k, a_m^*\} = \delta(k, m), \{a_k, a_m\} = 0 = \{a_k^*, a_m^*\}$$

Now define the 16 Dirac Gamma matrices

$$\gamma(k) = a_k + a_k^*, k = 1, 2, ..., 8, \gamma(k) = i(a_{k-8} - a_{k-8}^*), k = 9, 10, ..., 16$$

Clearly, they satisfy

$$\{\gamma(k), \gamma(m)\} = 2\delta(k, m), k, m = 1, 2, ..., 16$$

The matrices $J(k,m) = (1/4)[\gamma(k), \gamma(m)], 1 \le k < m \le 16$ satisfy the standard $SO(16)$-Lie algebra commutation relations. The generators $J(k,m), 1 \le k < m \le 16$ thus define a representation of the Lie algebra of $SO(16)$ in a 2^8 dimensional vector space. This representation is not irreducible. However, it can be decomposed into a direct sum of two irreducible representations of $SO(16)$, with each of these irreducible representations acting in a $2^7 = 128$ dimensional vector space. (For a nice proof of this, see V.S.Varadarajan, "Supersymmetry for mathematicians"). We denote the generators of this 128 dimensional irreducible Lie algebra by $(Q_k, k = 1, 2, ..., 128\}$. We can now define commutation relations $[X_i, Q_j]$ to be a linear combination of the $Q_k's$ in such a way that the generators $\{X_1, ..., X_{120}, Q_1, ..., Q_{128}\}$ span an irreducible Lie algebra of dimension $120 + 128 = 248$ which is precisely the dimension of the exceptional Lie algebra E_8. This Lie algebra plays a very important role in string theory. We have 10 bosonic field dimensions which are left propagators. Likewise we have 32 Fermionic field dimensions out of which 16 are left propagators and the other 16 are right propagators. On bosonization of the 32 Fermions, we obtain 16 boson dimensions which when added to the previous 10 give 26 bosonic dimensions which is precisely the critical dimension of the bosonic string required to guarantee a large number of zero norm states (See Green, Schwarz and Witten, "Superstring theory"). Now there are 32 Fermionic dimensions and these transform according to the $SO(32)$ group. We can partition these Fermions into two sets, the first comprising n Fermions that transform according to $SO(n)$ and the second comprising $32-n$ Fermions that transform according to $SO(32-n)$. Thus, the total 32 Fermions transform according to the group $SO(n) \times SO(32-n)$.

Remark: Suppose that we have field theory consisting of matter and gauge fields with a Lagrangian density locally invariant under the gauge group G. When the gauge symmetry of this Lagrangian is broken to a subgroup H by some of the bosons falling to the ground state, then some of the bosons become massless Goldstone Bosons which have only a kinetic component in their Lagrangian. When the local group transformation element $g(x)$ changes by a small amount to $g(x)(1+\epsilon\chi(x))$ where $\chi(x)$ is an element of the gauge group Lie algebra, then we can ask, by how much will the Lagrangian whose symmetry has been broken, deviate from the Lagrangian obtained by applying the local G-transformation $g(x)$? Further, if the Lagrangian whose symmetry has been broken to the subgroup H and if we apply a small perturbation of $h(x) \in H$ given by $h(x)(1+\epsilon\eta(x))$ where the Lie algebra element $\eta(x)$ is not in the Lie algebra of H, then by how much will the H-symmetric Lagrangian change by ?

9.4 Questions related to qualitative properties of quantum noise

[1] In a real quantum system, what are the sources of quantum noise? Two typical examples are (a) electrons executing thermal motion within a heated

resistor. The electrons are not in fact just point particles, they are spread out wave functions and hence the rapid thermal motion of the electrons should actually be described by rapid changes in their wave functions. This rapid change can be realized by adding quantum noise terms to the Hamiltonian of the system of electrons. Quantum noise in this context is therefore a mathematical model introduced to explain the rapid change in the wave functions of the thermally agitated electrons. Another example is that of an electron bound to its nucleus in an atom. A sudden impact by an external heavy particle on the nucleus can cause the potential generated by this nucleus to change rapidly thereby causing the electron to make a sudden random transition from a higher energy level to a lower energy level thereby releasing a photon. When this happens to several electrons, we get a random photon field which can be termed as Quantum Poisson noise. A third example is the light from a laser shining onto a system in which electrons are released from an source like an electron gun. The photons of the laser cause these incident electrons to get scattered and an output measurement of these scattered electrons as a function of time is precisely what we call Fermionic quantum noise. Likewise, the photons from the laser will also get scattered by the incident electrons and a measurement of these photons will yield quantum Bosonic noise. Photo-detection of electrons is another phenomenon involving quantum noise. Here, a stream of photons is incident on an array of atoms and by the photoelectric effect, electrons are released and when they hit a detector we generate current pulses which can be used to detect the presence of an incident photon. Each electron released corresponds to one photon. The energy of the released electrons equals $h\nu - W$ where ν is the frequency of the incident photon and W is the work function/binding energy of the electron to its nucleus. The number of released electrons which determines the strength of the electronic current equals the number of photons/intensity of the light which have frequency greater than W/h. The random hits of the electrons on the detector plate represent Fermionic noise which is in this case in fact generated by photonic ie, Bosonic noise. In general, when we have a very large aggregate of photons or electrons or more generally bosons and Fermions, then their effects on a quantum system can be described accurately by modeling the evolution of the quantum system in accordance with its Hamiltonian perturbed by quantum noise terms. Thus quantum noise theories are in fact mathematical models introduced to explain the effects of a large bath of elementary particles upon an quantum system like an atom.

[2] Is it possible to separate out the effects of thermal fluctuation and quantum noise in experimental realizations ? Thermal noise is usually modelled by a classical Gaussian process having Planckian black-body spectrum while quantum noise is modelled using non-commutative operator valued processes. Therefore, this question amounts to separating out the classical/purely commutative component from the non-commutiative component from a superposition of the two components. Mathematically, this can be achieved by commuting the process at two different times, so that the purely commutative/diagonal component is cancelled. Physically, we can use correlation properties to sep-

arate out the two. The classical thermal component usually has a long range correlations while the quantum component has a very short range correlations, thus if we predict the process based on its past samples with significant delay, the predictor output is likely to be only the classical component which can then be subtracted off just as in adaptive line enhancing techniques. Another way to separate out two such components is to take the correlations of the process in a given state. The first component may have a large correlation in a given state but not in the other while the second component has a large correlation in the latter state but not in the former. In such cases, by preparing the state appropriately, for measuring the process correlations, we can separate out the two components.

[3] Slight decrease in the nsr plot of the quantum filter. Actually, if the Hamiltonian of the noiseless system H_0 has eigenvalues $\omega(k), k = 1, 2, ..., p$, then the evolution operator is

$$U(t) = exp(-itH) = \sum_{k=1}^{p} exp(-i\omega(k)t)P_k$$

where P_k is the projection onto the eigenspace of H_0 with eigenvalue $\omega(k)$. Thus, $U(t)$ has oscillatory harmonic components. In the presence of quantum noise, these characteristic frequencies will still be present but will not be very prominent owing to the noise. The filtered output as well as the filtering error for an observable will have characteristic frequencies $\omega(k) - \omega(j), k < j$ which for a two state system amounts to just a single frequency $\omega(1) - \omega(2)$. Note that this is because a Heisenberg observable X evolves as

$$j_t(X) = U(t)^* XU(t) = \sum_{k,j} exp(i(\omega(k) - \omega(j))t)P_k X P_j$$

The dip shown in the graph is actually a low frequency oscillation which is to be explained as above.

[4] What is the reason for considering photon counting noise ? Typically most of the observed noises in nature are built out of the purely continuous Brownian motion and the purely discrete Poisson process. In quantum noise theory, we have correspondingly quantum Brownian motion which is attributed to several quantum particles moving along random trajectories and quantum Poisson process which is attributed to photons hitting a detector at random times or else photons causing electrons to get emitted from a metal which hit a detector at random times. The presence of a photo detector or an electron detector in our measurement apparatus is the precise reason for considering photon counting noise. It enables us to give a better model for the dynamics of photo-detection using the Hudson-Parthasarathy noisy Schrodinger equation. The statistics of photon counting noise is decided by the number of emitted photons while the statistics of quantum Brownian motion at the other extreme is decided by the nature of the quantum particle trajectory.

[5] Realistic systems in which quantum noise filtering theory can be applied. (a) To filter cavity resonator fields kept in the atmosphere containing noisy electrons, positrons and photons. (b) To estimate the spin of an electron based on its interaction with a magnetic field when the atom having the electron is inside a bath. (c) To probe into the ultimate structure of matter by allowing a system containing elementary particles to interact with elementary particles coming from a source and studying the scattering, absorption and emission amplitudes of the resulting particles using measurement apparatus which are subject to measurement noise.

[6] If $X(r)$ is a quantum field within a cavity with the initial state of the field being $\rho(\theta)$ depending upon a parameter vector θ, and the state evolves according to the Schrodinger dynamics with Hamiltonian H we take measurements of the field at times $t_1, ..., t_N$ at the points $r_1, ..., r_N$ respectively each time noting the measurement outcome and incoroprating the collapse postulate for the state. Then calculate the joint probabilities for the measured outcome. If the state evolves according to the GKSL master equation for open quantum systems, then do the same. Note that we have a spectral decomposition

$$X(r) = \sum_m \lambda_m(r)|e_m(r)><e_m(r)|$$

for the quantum field at each spatial point r. The PVM used to measure the field at r is $\{P_m(r) = |e_m(r)><e_m(r)| : m = 1, 2, ...\}$. If ρ_0 is the state of the field just before $X(r)$ is measured and ρ_1 just after it is measured, then

$$\rho_1 = \sum_m P_m(r)\rho_0 P_m(r)$$

if the measurement outcome is not noted while if the measurement outcome is noted to be m, then

$$\rho_1 = P_m(r)\rho_0 P_m(r)/Tr(\rho_0 P_m(r))$$

[7] Now suppose that the Lindblad noise operators of an open quantum system are small, ie, they have a small multiplicative factor of $\sqrt{\epsilon}$. Take two distinct eigen-states $|n>, |m>$ of the Hamiltonian. If the Lindblad parameter is zero, then there is zero probability of the system making a transition from one state to the other. when this parameter is small and non-zero, calcluate the probability of the system making a transition and evaluate the rate at which this probability goes to zero as $\epsilon \to 0$.

[8] Consider an atom with an electron placed within a cavity resonator box of arbitrary shape. Let V denote the volume region of the cavity and let $S = \partial V$ denote its boundary surface. The electromagnetic fields within this cavity satisfy the three dimensional Helmholtz equation

$$(\nabla^2 + \omega^2\mu\epsilon)(\mathbf{E}, \mathbf{H}(\omega, r)) = 0$$

These fields also satisfy the boundary conditions that the tangential components of **E** and the normal component of **H** vanishes on the S. Further, the components of these fields are related by the Maxwell curl equations

$$curl\mathbf{E} = -j\omega\mu\mathbf{H}, curl\mathbf{H} = j\omega\epsilon\mathbf{E}$$

In fact, these two curl equations imply the above Helmholtz equation. Choose an orthogonal system of coordinates (q_1, q_2, q_3) such that $q_1 = c$ corresponds to the cavity boundary S. Writing the curl and ∇^2 equations in this curvilinear system using the Lames' coefficients as

$$det\begin{pmatrix} e_1/h_2h_3 & e_2/h_3h_1 & e_3/h_1h_2 \\ \partial_1 & \partial_2 & \partial_3 \\ h_1E_1 & h_2E_2 & h_3E_3 \end{pmatrix} = -j\omega\mu(H_1e_1 + H_2e_2 + H_3e_3)$$

and likewise

$$det\begin{pmatrix} e_1/h_2h_3 & e_2/h_3h_1 & e_3/h_1h_2 \\ \partial_1 & \partial_2 & \partial_3 \\ h_1H_1 & h_2H_2 & h_3H_3 \end{pmatrix} = j\omega\epsilon(E_1e_1 + E_2e_2 + E_3e_3)$$

The boundary conditions translate to $E_2 = E_3 = 0, H_1 = 0$ at $q_1 = c$. The Helmholtz equation, the first two components of the Maxwell curl equations and the boundary conditions imply that the general solution to the fields has the form

$$\mathbf{E}(t, \mathbf{q}) = \sum_n Re(c(n).exp(j\omega(n)t)\psi_n(\mathbf{q})$$

$$\mathbf{H}(t, \mathbf{q}) = \sum_n Re(d(n).exp(j\omega(n)'t)\phi_n(\mathbf{q})$$

where $\psi_n(\mathbf{q})$ and $\phi_n(\mathbf{q})$ are 3-vector valued real eigenfunctions of ∇^2 with eigenvalues $-\omega(n)^2\epsilon\mu$ and $-\omega(n)'^2\epsilon\mu$ satisfying three first two Maxwell curl equations and the boundary conditions that the last two components of $\psi_n(\mathbf{q})$ vanish on S while the first component of $\psi_n(\mathbf{q})$ vanishes on S.

9.5 Questions on Pattern Recognition

[1] [a] Consider N iid random vectors $\mathbf{X}_n, n = 1, 2, ..., N$ where $\mathbf{X}_n$ has the probability density $\sum_{k=1}^{K} \pi(k)N(\mathbf{x}|\mu_k, \Sigma_k)$, ie, a Gaussian mixture density. Derive the optimal ML equations for estimating the parameters $\theta = \{\pi(k), \mu_k, \sigma_k) : k = 1, 2, ..., K\}$ from the measurements $\mathbf{X} = \{\mathbf{X}_n : n = 1, 2, ..., N\}$ by maximizing

$$ln(p(\mathbf{X}|\theta)) = \sum_{n=1}^{N} ln(\sum_{k=1}^{K} \pi(k)N(\mathbf{X}_n|\mu_k, \Sigma_k))$$

[b] By introducing latent random vectors $\mathbf{z}(n) = \{z(n,k) : k = 1,2,...,K\}, n$ 1, 2, ..., N which are iid with $P(z(n,k) = 1) = \pi(k)$ such that for each n, exactl one of the $z(n,k)'s$ equals one and the others are zero, cast the ML paramete estimation equation in the recursive EM form, namely as

$$\theta_{m+1} = argmax_\theta Q(\theta, \theta_m)$$

where

$$Q(\theta, \theta_m) = \sum_{\mathbf{Z}} ln(p(\mathbf{X}, \mathbf{Z}|\theta)).p(\mathbf{Z}|\mathbf{X}, \theta_m)$$

with $\mathbf{Z} = \{\mathbf{z}(n) : n = 1, 2, ..., N\}$. Specifically, show that

$$Q(\theta, \theta_m) = \sum_{n,k} \gamma_{nk}(\mathbf{X}, \theta_m) ln(\pi(k) N(\mathbf{X}_n|\mu(k), \Sigma_k))) p(\mathbf{Z}|\mathbf{X}, \theta_m)$$

where

$$\gamma_{n,k}(\mathbf{X}, \theta_m) = \mathbb{E}(z(n,k)|\mathbf{X}, \theta_m)$$

$$= \sum_{\mathbf{Z}} z(n,k) p(\mathbf{Z}|\mathbf{X}, \theta_m) = \sum_{z(n,k)} z(n,k) p(z(n,k)|\mathbf{X}, \theta_m)$$

Derive a formula for $p(z(n,k)|\mathbf{X}, \theta_m)$. Derive the optimal equations for θ_{m+} by maximizing $Q(\theta, \theta_m)$.

[2] Let $\mathbf{X}_n, n = 1, 2, ..., N$ be vectors in $\mathbb{R}^N$. Choose $K << N$ and vector $\{\mu(k) : k = 1, 2, ..., K\}$ and a partition

$$\mathcal{P} = E_1 \cup E_2 \cup ... \cup E_K$$

of $\{1, 2, ..., N\}$ into K disjoint non-empty subsets such that

$$\mathcal{E} = \sum_{k=1}^{K} \sum_{n \in E_k} \| \mathbf{X}_n - \mu(k) \|^2$$

is a minimum. Interpret this "Clustering principle" from the viewpoint of dat compression and image segmentation. Explain how you would obtain an itera tive solution to this clustering problem by defining

$$r(n,k) = 1, n \in E_k, r(n,k) = 0, n \notin E_k$$

and expressing

$$\mathcal{E} = \sum_{n,k} r(n,k) \| \mathbf{X}_n - \mu(k) \|^2$$

Relate this clustering problem to the problem of calculating the $\mu(k)'s$ and th $\pi(k)'s$ in the Gaussian mixture model. Specifically, compare the equations fo calculating $r(n,k), \mu(k)$ in this clustering problem to the equations for calcu lating the same in the GMM. Explain how the GMM model is a "smoothene version" of the clustering model by remarking that the event $\mathbf{X}_n \in E_k$ is to be interpreted in the GMM context by saying that $\mathbf{X}_n$ has the $N(\mathbf{x}|\mu(k), \Sigma_k)$ distribution and that $\sum_{n=1}^{N} r(n,k)/N = card(E_k)/N$ corresponds to the probability $\pi(k)$ of the kth component in the GMM occurring. Derive the explicit value that $r(n,k)$ corresponds to in the GMM by showing that the optimal $\pi(k)$ in the GMM satisfies

$$\pi(k) = \sum_{n=1}^{N} \frac{\pi(k)N(\mathbf{X}_n|\mu_k, \Sigma_k)}{\sum_{m=1}^{K} \pi(m)N(\mathbf{X}_n|\mu(m), \Sigma_m)}$$

Compare this with the quantity $\pi(k) = \sum_{n=1}^{N}(r(n,k)/N), \sum_{n,k} r(n,k) = N$ in the clustering problem.

[3] Let $\{Z(n) : n = 1, 2, ..., N\}$ be a Markov chain in discrete state space and let $\pi(Z(n), Z(n+1)|\theta)$ be its one step transition probability distribution. Let $\pi_0(Z(0))$ denote its initial distribution. Let $\mathbf{X} = (\mathbf{X}(n) : n = 0, 1, ..., N)$ be the HMM emission random variables with the conditional probability density $p(X(n)|Z(n), \theta)$ defined.

[a] Show that the EM algorithm for estimating θ recursively amounts to maximizing $Q(\theta, \theta_0)$ where

$$Q(\theta, \theta_0) =$$

$$\sum_{n=0}^{N} \mathbb{E}(ln(p(X(n)|Z(n), \theta))|\mathbf{X}, \theta_0)$$

$$+ \sum_{n=0}^{N-1} \mathbb{E}(ln(\pi(Z(n), Z(n+1)|\theta_0))|\mathbf{X}, \theta_0)$$

[b] Show that for evaluating the conditional expectation in [a], we require the conditional probabilities $p(Z(n)|\mathbf{X}, \theta_0)$ and $p(Z(n), Z(n+1)|\mathbf{X}, \theta_0)$. Explain how you would evaluate these conditional probabilities recursively by first proving the relations

$$p(\mathbf{X}|Z(n), \theta_0) = p(X(N), X(N-1), ..., X(n)|Z(n), \theta_0).p(X(n-1), X(n-2), ..., X(0)|Z(n), \theta_0)$$

and

$$p(\mathbf{X}|Z(n+1), Z(n), \theta_0) =$$

$$p(X(N), ..., X(n+1)|Z(n+1, \theta_0).p(X(n-1), ..., X(0)|Z(n), \theta_0)$$

4 [a] How would you approximate (based on matching only first and second order statistics) an N-dimensional random vector $\mathbf{X}$ by a $p << N$ dimensional random vector $\mathbf{S}$ using the linear model

$$\mathbf{X} \approx \mathbf{WS} + \mu + \mathbf{V} = \hat{\mathbf{X}} + \mathbf{V}, \hat{\mathbf{X}} = \mathbf{WS} + \mu$$

where $\mathbf{S}, \mathbf{V}$ are zero mean mutually uncorrelated random vectors with covariances Σ_S, Σ_V respectively and $\mathbf{W}$ is an $N \times p$ matrix to be determined. To solve this problem, you must assume that $\mathbb{E}(\mathbf{X})$ is well approximated by μ and $Cov(\mathbf{X})$ is well approximated by $\mathbf{W}\Sigma_S\mathbf{W}^T + \Sigma_V$.

[b] Now consider the maximum likelihood version of the above PCA in which the data is modeled by an iid sequence:

$$\mathbf{X}_n = \mathbf{S}_n + \mu + \mathbf{V}_n, n = 1, 2, ..., N$$

where the $\mathbf{S}_n's$ are iid $N(\mathbf{0}, \Sigma_S)$ random vectors, the $\mathbf{V}_n's$ are iid $N(\mathbf{0}, \Sigma_v)$ random vectors with the two sequences being mutually independent. Calculate the equations that determine the optimal choice of $\mathbf{W}, \mu$ that would maximize $p(\mathbf{X}_n, n = 1, 2, ..., N|\mathbf{W}, \mu)$. Also by assuming the $\mathbf{S}_n's$ to be latent random vectors, derive the EM algorithm recursive equations for estimating $\mathbf{W}, \mu$ from the $\mathbf{X}_n, n = 1, 2, ..., N$.

[c] If $\Sigma_V = \sigma^2\mathbf{I}$, then derive the EM recursions for estimating $\mathbf{W}, \mu, \Sigma_S, \sigma^2$ and compare with the exact maximum likelihood equations. Explain why this process is called principle component analysis and explain how much data compression we have achieved by this process.

[d] Consider now the following data matrix version of PCA. $\mathbf{X}$ is a given $N \times M$ real matrix. Show that it has a singular value decomposition (SVD)

$$\mathbf{X} = \mathbf{U}\Sigma\mathbf{V}$$

where $\mathbf{U}$ and $\mathbf{V}$ are respectively $N \times N$ and $M \times M$ real orthogonal matrices and Σ is a diagonal $N \times M$ matrix with positive diagonal entries either N or M in number, depending on whether N or M is smaller. Let $r < min(N, M)$ and then using the SVD, construct an $N \times M$ matrix $\mathbf{Y}$ of rank r so that $\| \mathbf{X} - \mathbf{Y} \|_F^2$ is a minimum where $\| Z \|_F = \sqrt{Tr(\mathbf{Z}^T\mathbf{Z})}$ is the Frobenius norm of the real matrix $\mathbf{Z}$.

[5] Consider the binary hypothesis testing problem in which under the hypothesis H_1, each sample of the iid sequence $X(n), n \geq 1$ has the probability density $p_1(X)$ while under H_0, each sample of this iid sequence has the probability density $p_0(X)$. Show that the Neyman-Pearson criterion of decision based on the observations $X(n), n = 1, 2, ..., N$ leads to the test of deciding H_1 in the case when $N^{-1}\sum_{n=1}^{N} L(X(n)) \geq \eta$ and deciding H_0 in the case when $N^{-1}\sum_{n=1}^{N} L(X(n)) < \eta$ where

$$L(X) = ln(p_1(X)/p_0(X))$$

and the threshold $\eta(N)$ is such that

$$0 < \alpha = \int_{Z_1} \Pi_{n=1}^{N}(p_0(X(n))dX(n)) = P(N^{-1}\sum_{n=1}^{N} L(X(n)) > \eta | H_0)$$

Here, Z_1 is the region where H_1 is decided. Estimate this probability by the LDP and calculate using the LDP the value of η for given α. Hence use this threshold η to calculate the other error probability

$$\beta(N) = \int_{Z_1^c} \Pi_{n=1}^{N}(p_1(X(n))dX(n)) = P(N^{-1}\sum_{n=1}^{N} L(X(n)) < \eta | H_1)$$

and show that for any $\alpha > 0$ however small, $\beta(N)$ if $\eta = \eta(N)$ is obtained as above for the given α, then $\beta(N)$ converges to zero at the rate of $D(p_0|p_1)$, ie,

$$N^{-1}log(\beta(N)) \to D(p_0|p_1), N \to \infty$$

where

$$D(p_0|p_1) = \int p_0(X).log(p_0(X)/p_1(X)).dX$$

[6] This problem deals with non-parametric density estimation. Let $p_1(X), p_0(X)$ be two different probability densities of a random vector $X \in \mathbb{R}^n$. To estimate p_1 at a given point X, we take samples from this distribution and draw a sphere of minimum radius R_1 with centre at X, for which the first sample point falls within this sphere. Let $V(R_1)$ denote the volume of such a sphere. Then, estimate $p_1(X)$ as $1/V(R_1)$. Likewise do so for estimating $p_0(X)$. Let the minimum radius for this be R_0. Justify using the law of large numbers that such a non-parametric density estimate is reliable. Show that given a sample point X, in order to test whether X came from p_1 or from p_0, we use the ML criterion, namely decide p_1 if $\hat{p}_1(X) \geq \hat{p}_0(X)$ and decide p_0 otherwise, where $\hat{p}$ denotes the estimate of p obtained using the above scheme. Show that this MLE method is the same as the minimum distance method, in the sense that if the first sample point that we decide p_1 if $R_1(X) < R_0(X)$ and decide p_0 otherwise where $R_1(X)$ is the distance of the first point taken from the p_1 sample from X and likewise $R_0(X)$ is the distance of the first point taken from the p_0 sample from X.

9.6 Appendix

Aim of this section: To calculate the rate functions for the response of systems described by linear and nonlinear ode's and pde's driven by weak Gaussian and Non-Gaussian noise with the system having control parameters and then to design these control parameters so that the probability that the system response deviates from the noiseless response for a given set of the control parameters by an amount greater than a given small threshold is minimized.

[1] Parametrize $SU(2,\mathbb{C})$ by a set of three real parameters and determine the differential equations for a function on $SL(2,\mathbb{C})$ to be invariant under the map $g \to hgh^{-1}$ for $g, h \in SL(2,\mathbb{C})$. These differential equations should be expressed in terms of the three real parameters chosen in our parametrization of $SU(2,\mathbb{C})$

[2] Let $\mathfrak{g}$ be a Lie algebra. Let ρ be a nil-representation of this Lie algebra in a vector space V. Show that there exists a non-zero vector $v \in V$ such that $\rho(X)v = 0 \forall X \in \mathfrak{g}$. Hence deduce that there exist ρ-invariant subspaces $0 = V_0 \subset V_1 \subset ... \subset V_p = V$ such that

$$V_{i+1} = \{v \in V : \rho(X)v \in V_i\}, i = 0, 1, ..., p-1$$

By applying this result to the adjoint reperesentation, deduce that any nilpotent Lie algebra is also solvable.

Transfer learning algorithm for learning about the functions that relate speech parameters to EEG parameters.

[71] String theoretic corrections to classical field theory Lagrangians: Let $X^\mu(\tau, \sigma)$ be a quantum Bosonic string in D dimensions. It satisfies the wave equation

$$\partial_\tau^2 X^\mu - \partial_\sigma^2 X^\mu = 0$$

and the solution to this equation corresponding to a right moving wave is given by

$$X^\mu(\tau, \sigma) = x^\mu + p^\mu \tau + \sum_{n \neq 0} \alpha^\mu(n) exp(in(\tau - \sigma))/n$$

where by virtue of the canonical commutation relations

$$[X^\mu(\tau, \sigma), \partial_\tau X^\nu(\tau, \sigma')] = \eta^{\mu\nu}\delta(\sigma - \sigma')$$

and hence

$$[\alpha^\mu(m), \alpha^\nu(n)] = \eta^{\mu\nu}.n.\delta[m + n]$$

Now consider the string action in a curved space-time given by

$$S_1(g, X) = (1/2)\int g_{\mu\nu}(X(\tau, \sigma))\sqrt{-h(\tau, \sigma)}h^{\alpha\beta}(\tau, \sigma)\partial_\alpha X^\mu.\partial_\beta X^\nu d\tau.d\sigma$$

where $g_{\mu\nu}(X)$ is the metric of D-dimensional space-time and $h_{\alpha\beta}$ is the metric of the two dimensional string world sheet. Let $S_2(g)$ denote the D-dimensional Einstein-Hilbert action:

$$S_2(g) = K.\int R(X)\sqrt{-g(X)}d^D X$$

where $R(X)$ is the scalar curvature of the D-dimensional space-time manifold. Define $S_3[g]$ by the equation

$$exp(iS_3(g)/h) = K.\int exp(i(S_1(g,X)/h))DX$$

Then $S_3(g)$ can be regarded as a string theoretic quantum correction to the Einstein-Hilbert action $S_2(g)$. We can expand $S_3[g]$ in powers of Planck's constant h and retain terms upto a given power of it. The question is "Is there and operator theoretic formalism to derive these string theoretic quantum corrections to the Einstein-Hilbert action ?"

To this end, we write

$$X^\mu(\tau,\sigma) = x^\mu + \delta X^\mu(\tau,\sigma)$$

and evaluate the quantum average of the string action:

$$< \int g_{\mu\nu}(x + \delta X(\tau,\sigma))h^{\alpha\beta}\sqrt{-h}\partial_\alpha \delta X^\mu(\tau,\sigma)\partial_\beta \delta X^\nu(\tau,\sigma)d\tau.\sigma >$$

where the average may for example be taken in vacuum or more generally in a coherent state for the annihilation-creation operators $\{\alpha^\mu(n)\}$. This results in correction to the Einstein-Hilbert action. Actually, it should be borne in mind that in a curved space-time, the quantum string field will actually depend upon the metric $g_{\mu\nu}$ and hence the corrections to the Einstein-Hilbert action will contain nonlinear terms in the metric and its partial derivatives.

Consider for example an equation for the string field dynamics defined by the action

$$S[X] = \int (1/2)\eta^{\alpha\beta}X^\mu_{,\alpha}X_{\mu,\beta}d^2\sigma + \int B_{\mu\nu}(X)\epsilon^{\alpha\beta}X^\mu_{,\alpha}X_{\mu,\beta}d^2\sigma$$

The equations of motion obtained from the variational principle $\delta S[X] = 0$ can be derived using

$$\delta[B_{\mu\nu}(X)\epsilon^{\alpha\beta}X^\mu_{,\alpha}X^\nu_{,\beta}] =$$

$$B_{\mu\nu,\rho}(X)\epsilon^{\alpha\beta}X^\mu_{,\alpha}X^\nu_{,\beta}\delta X^\rho - 2(B_{\mu\nu}(X)\epsilon^{\alpha\beta}X^\nu_{,\beta})_{,\alpha}\delta X^\mu$$

$$= \epsilon^{\alpha\beta}[B_{\rho\nu,\mu}X^\rho_{,\alpha}X^\nu_{,\beta} - 2B_{\mu\nu,\rho}\epsilon^{\alpha\beta}X^\rho_{,\alpha}X^\nu_{,\beta}]\delta X^\mu$$

$$= H_{\rho\nu\mu}(X)\epsilon^{\alpha\beta}X^\rho_{,\alpha}X^\nu_{,\beta}\delta X^\mu$$

where

$$H_{\rho\nu\mu} = B_{\rho\nu,\mu} + B_{\nu\mu,\rho} + B_{\mu\rho nu}$$

is the totally antisymmetric field tensor derived from the antisymmetric gauge potential tensor $B_{\mu\nu}$. The string field equations are therefore

$$\Box X_\mu = K.H_{\mu\rho\nu}(X)dX^\rho \wedge dX^\nu = K.H_{\mu\rho\nu}(X)\epsilon^{\alpha\beta}X^\rho_{,\alpha}X^\nu_{,\beta}$$

Using perturbation theory applied to this equation, we can now derive the ap-proximate correction to the string field propagator produced by the presence o the background gauge field.

The LDP problem: Given that the gauge field $B_{\mu\nu}(X)$ and the metric fielc $g_{\mu\nu}(X)$ are small random field perturbations of the zero gauge and the fla space-time metric, compute the rate function of the string field and hence the probability that this field will deviate from the flat space-time and zero gauge field solution by an amount greater than a given threshold. The second problen is to include a control gauge field term and a control metric field term speci-fied completely except for a set of p unspecified control parameters θ and ther when these two fields undergo small random perturbations, to compute the rate function for the string field in terms of the control parameters and to adjus these control parameters so that the probability of the string field deviating from a given specified non-random string field by an amount greater than a given threshold ϵ is a minimum.

Chapter 10

LDP in Markov Chain and Queueing Theory with Quantum Mechanical Applications

10.1 Notes on applications of LDP to stochastic processes and queueing theory

[1] The general $G_1/G_2/1$ queue: Let arrivals take place at times $X_n, n = 1, 2, ...$ in a single server queue. Let the waiting time for the nth customer who arrives at time X_n to commence service be W_n and let his service time be Y_n. Given the sequence $\{X_n, Y_n : n \geq 1\}$, the first problem is to determine a recursive relationship between W_{n+1} and W_n.

It is clear that the nth customer departs at the time $X_n + W_n + Y_n$. Thus, if $X_{n+1} < X_n + W_n + Y_n$, then $W_{n+1} = X_n + W_n + Y_n - X_{n+1}$ while if $X_{n+1} > X_n + W_n + Y_n$, then $W_{n+1} = 0$. In other words, we have derived the recursion

$$W_{n+1} = max(0, X_n - X_{n+1} + W_n + Y_n)$$

The next problem is to determine the length $Q(t)$ of the queue at time t in terms of the sequence $\{X_n, Y_n : n \geq 1\}$. It is clear that the nth customer is in the queue at time t iff $X_n \leq t$ and $D_n = X_n + W_n + Y_n > t$, ie, iff he arrives before time t and departs after time t. Here, D_n is the departure time of the nth customer. Thus, we get

$$Q(t, \omega) = \sum_{n \geq 1} \chi_{X_n \leq t < D_n}(\omega)$$

From this we get

$$T^{-1}\int_0^T Q(t)dt = T^{-1}\sum_{n\geq 1}(min(T, D_n) - min(T, X_n))$$

Note that the r.v's X_n, Y_n and hence W_n are all non-negative and hence $X_n \leq D_n$. Now, let $N(T)$ denote the number of arrivals in the time interval $[0, T]$. Thus,

$$N(T) = max(n : X_n \leq T)$$

Equivalently,

$$\{N(T) = n\} = \{X_n \leq T < X_{n+1}\}$$

We get

$$T^{-1}\int_0^T Q(t)dt = (\frac{N(T)}{T}).N(T)^{-1}.\sum_{n=1}^{N(T)}(min(T, D_n) - X_n)$$

and it follows that if

$$lim_{T\to\infty}\frac{N(T)}{T} = \lambda$$

the mean arrival rate and if

$$lim_{N\to\infty}N^{-1}\sum_{n=1}^{N}(D_n - X_n) = lim_{N\to\infty}N^{-1}\sum_{n=1}^{N}(W_n + Y_n) = \mathbb{E}(W + Y)$$

where $\mathbb{E}(W)$ is the mean waiting time for a customer when the queue has attained a steady/equilibrium state, then

$$lim_{T\to\infty}T^{-1}\int_0^T Q(t)dt = \lambda.\mathbb{E}(W + Y)$$

where $Y = Y_1$. Here, we are assuming that the $X_n's$ are identically distributed and so are the $Y_n's$. We are not putting the restriction of independence. Note that $\mathbb{E}(W + T)$ is the mean sojourn time for a customer at equilibrium, ie, ie the mean time that a customer spends in the queue after his arrival. This result is known as Little's theorem.

Problem:

Virtual waiting time at a queue: The virtual waiting time $W(t)$ just after time t is the time taken by a customer if he arrives at time t for his service to commence. The last customer who arrives in the queue at or before time t is the $N(t)^{th}$ customer. $W(t)$ is clearly the time taken for the completion of the service of this customer. Thus,

$$W(t) = W_{N(t)} + Y_{N(t)}$$

10.2 LDP problems in Markov chain theory

Let $X(n), n = 0, 1, 2, ...$ be a Markov chain in discrete time with transition probability distribution $\pi(x, y)$ and discrete state space E. The rate function for empirical measures

$$\mu_N = N^{-1}.\sum_{n=1}^{N} \delta_{X(n)}, N \geq 1$$

is known to be given by

$$I(q) = \sup_{u>0} \sum_{k \in E} q(k) log(u(k)/(\pi u)(k))$$

Now suppose that the transition probability $\pi(x, y) = \pi(x, y|\theta)$ is parametrized by a control parameter vector θ. Further assume that some noise is present in this Markov chain so that the noise perturbed process is another Markov process with transition probability distribution $\rho(x, y|\theta)$. The original parameter for the noiseless chain is θ_0 and let $p_0(x)$ be a stationary distribution for this chain:

$$\sum_x p_0(x)\pi(x, y|\theta_0) = p_0(y)$$

or equivalently, in matrix notation,

$$p_0^T \pi = p_0$$

The rate function for the empirical measures of the perturbed chain is

$$J(q|\theta) = sup_{u>0} \sum_x q(x) ln(u(x)/\rho u(x))$$

$$= sup_{u>0} \sum_x q(x) ln(u(x)/(\sum_y \rho(x, y|\theta)u(y))$$

The aim is to choose the parameter θ so that the probability that this empirical measure will deviate from p_0 by an amount greater than a threshold ϵ is minimized.

10.3 Continuity and non-differentiability of the Brownian sample paths

Let $h = 2/2^n$.

$$P(|W((i+2)h) - W((i+1)h)|/h, |W((i+1)h) - W(ih)|/h, |W(ih) - W((i-1)h)|/h < K)$$

$$= (\Phi(K.\sqrt{h}) - \Phi(-K.\sqrt{h}))^3 \leq (2K.sqrth/2\pi)^3$$

10.4 Renewal processes in quantum mechanics

Let $X(n), n = 1, 2, ...$ be non-negative iid random variables and define $S(n) = X(1) + ... + X(n), n \geq 1, S(0) = 0$. Define for $t \geq 0$

$$N(t) = max(n : S(n) \leq t)$$

Define a signal that jumps by random independent amounts at the renewal epochs $S(n)$:

$$\xi(t) = \sum_{k=1}^{N(t)} Y(k), t \geq 0$$

where the $Y(k)'s$ are iid r.v.s independent of the $X(n)'s$. Note that

$$d\xi(t) = \xi(t + dt) - \xi(t-) = Y(N(t) + 1)dN(t)$$

It follows that

$$\int_0^T f(t)d\xi(t) = \sum_{n=0}^{N(T)} f(S(n))Y(n + 1)$$

The first problem is to compute the moment generating functional of the process $\xi(t), t \geq 0$:

$$M_\xi(f) = \mathbb{E}[exp(\int_0^T f(t)d\xi(t))]$$

The next problem is to look at the following application of the process $\xi(t)$. Each time that a renewal epoch occurs, an electron hits a plate generating a random current. The total current generated upto time t can then be modeled as $\xi(t)$. An alternative model for this current is to assume that a pulse $h(t)$ is generated in the plate when the electron hits the plate at time $t = 0$. It follows that when the electron hits the plate at time $S(n)$, the current generated is $h(t - S(n))$. We can also incorporate a random parameter, ie, noise in each of these current pulses so that the total current generated in the plate at time t is given by

$$I(t) = \sum_n h(t - S(n), Y(n))$$

If we assume that $h(t) = 0$ for $t < 0$, then we can equivalently write the above expression as

$$I(t) = \sum_{n=0}^{N(t)} h(t - S(n), Y(n))$$

This current $I(t)$ generates a voltage when passed through a resistor and this voltage generates an electric field between two conducting plates. A charge attached to a quantum harmonic oscillator interacts with this electric field and hence the Hamiltonian of this randomly forced oscillator becomes

$$H(t) = (q^2 + p^2)/2 - I(t)q$$

The problem is to calculate the average transition probabilities for this forced oscillator and under the condition that the current is a "weak" random process, to evaluate the large deviation rate function for the transition probabilities.

Large deviation principle and the Atiyah-Singer Index Theorem: Consider a non-Abelian gauge connection $A_\mu(x)$ and the associated covariant derivative

$$\nabla_\mu = \partial_\mu + ieA_\mu$$

Let $\gamma^\mu(x) = \gamma^a e_a^\mu(x)$ denote the local Dirac Gamma matrices relative to the tetrad $e_a^\mu(x)$. Note that

$$\{\gamma^\mu(x), \gamma^\nu(x)\} = \{gamma^a, \gamma^b\} e_a^\mu(x) e_b^\nu(x) = \eta^{ab} e_a^\mu(x) e_b^\nu(x) = g^{\mu\nu}(x)$$

Assume the background metric and hence the tetrad to be a randomly fluctuating quantity around a non-random metric.

Chapter 11

LDP in Device Physics, Quantum Scattering Amplitudes, Quantum Filtering and Quantum Antennas

11.1 Large deviations in vacuum polarization

A photon arrives, it produces and electron-positron pair and this pair again combines to produce a photon. The electron-positron loop determines the amplitude of this one loop process. Let μ denote the polarization of the incoming photon and ν that of the outgoing photon. Then the amplitude for this process is given by

$$\Pi^{\mu\nu}(q) = \int Tr[\gamma^\mu S(p)\gamma^\nu S(p-q)]d^4p$$

where $S(p)$ is the electron propagator:

$$S(p) = [\gamma.p - m + i0]^{-1}$$

Now consider the case when we do not have just one electron and one positron in this polarization process but a whole distribution of particles (In the language of Feynman, partons), then the electron propagator has to be replaced by an integral

$$S(p) = \int [\gamma.p - m + i0]^{-1} d\mu(m)$$

where $\mu(.)$ is a measure in the space of the parton masses. Now, it is very likely that the parton distribution $d\mu(.)$ will be concentrated around the electron mass and so we can equivalently specify it by a large deviation rate function $I(m)$: $\mu(dm) = d\mu(m) = exp(-I(m)/\epsilon)$. The aim is to derive a formula for this rate function by measuring amplitudes that involve the vacuum polarization. For example, the vacuum polarization gives us a one loop correction to the electron propagator as

$$S(p)_c = S(p) - S(p)\Pi(p)S(p)$$

The corresponding shift in the electron self energy is then given by

11.2 An application of the EKF and LDP to estimating the current in a pn junction

The current density consists of a drift component and a diffusion component:

$$J(t,x) = J_{dr}(t,x) + J_{diff}(t,x)$$

where

$$J_{dr}(t,x) = \mu.E(t,x), J_{diff}(t,x) = D.\partial n(t,x)/\partial x$$

with $n(t,x)$ denoting the electronic concentration. μ is the electron mobility and $E(t,x)$ is the electric field. The potential difference between the ends of the semiconductor is $\Delta V(t)$. Thus if $V(t,x)$ is the potential within the semiconductor body and if $x = 0$ and $x = L$ are the end terminals of the semiconductor, then

$$\Delta V(t) = V(t,0) - V(t,L)$$

and we have the Poisson equation

$$\partial^2 V(t,x)/\partial x^2 = n(t,x)/\epsilon, E(t,x) = -\partial V(t,x)/\partial x$$

Further, assuming no generation of electrons, we have the current conservation equation

$$\partial J(t,x)/\partial x + \partial n(t,x)/\partial t = 0$$

From these equations, we easily derive

$$-mu\partial V(t,x)/\partial x^2 + D\partial^2 n(t,x)/\partial x^2 + \partial n(t,x)/\partial t = 0$$

or equivalently,

$$-\mu n(t,x)/\epsilon + D\partial^2 n(t,x)/\partial x^2 + \partial n(t,x)/\partial t = 0$$

Along with the terminal endpoint conditions on the potential, these equations determine the current as a function of the potential difference between the two terminals. If the potential difference is a constant ΔV, then the steady state concentration $n(x)$ satisfies the differential equation

$$-\mu n(x)/\epsilon + Dn''(x) = 0$$

and from basic thermodynamics, if $V(x)$ is the equilibrium potential distribution within the semiconductor, then

$$n(x) = C.exp(eV(x)/kT)$$

and further,

$$-\mu V''(x) + Dn''(x) = 0$$

which solves to give

$$V(x) = Dn(x)/\mu + ax + b, n(x) = (\mu/D)(V(x) - ax - b)$$

Thus, the equilibrium potential satisfies

$$V''(x) = (C/\epsilon)exp(eV(x)/kT) = (\mu/D\epsilon)(V(x) - ax - b)$$

from which, we can derive a formula for the diffusion coefficient D in terms of the mobility μ and temperature T or vice versa.

The LDP problem: Suppose that there are random fluctuations in the potential difference $\Delta V(t)$, say the potential difference

$$\Delta V(t) = \Delta V_0(t) + \sqrt{\epsilon}w(t)$$

where $w(t)$ is standard white Gaussian noise WGN or equivalently the derivative of standard Brownian motion. Then, calculate the rate function of the current $J(t,x)$ in the semiconductor and the current at the terminals $J(t,0), J(t,L)$.

11.3 Large deviation problems in quantum stochastic filtering theory

11.4 Large deviation problems in quantum antennas

Dirac field and Maxwell field in the presence of external classical random current and radiation fields. When these external fields are of small amplitude, then the objective is to compute the rate function for the quantum average of the radiation fields in the far field zone. The Dirac and Maxwell equations for the second quantized wave functions are given by

$$(i\gamma.\partial - m)\psi(x) = -e\gamma.A_q(x).\psi(x) - e\gamma.A_c(x).\psi(x),$$

$$\Box A_q^\mu(x) = -e\bar{\psi}(x)\gamma^\mu\psi(x) + J_c^\mu(x)$$

where $J_c^\mu(x)$ is the classical random electromagnetic field applied externally through a laser while $J_c^\mu(x)$ is the external classical random current density field applied through a probe. We now define the electron propagator

$$S_{lm}(x,y) =< 0|T(\psi_l(x)\psi_m(y)^*)|0 >= \theta(x^0-y^0) < 0|\psi_l(x)\psi_m(y)^*|0 > -\theta(y^0-x^0)$$

$$< 0|\psi_m(y)^*\psi_l(x)|0 >$$

Note that

$$\bar{\psi}(x) = \psi(x)^*\gamma^0$$

and find using the canonical equal time anticommutation relations

$$\{\psi_l(t,r), \psi_m(t,r')^*\} = \delta^3(r-r')$$

that

$$(i\gamma.\partial - m)S_{lm}(x,y) = \delta_{lm}\delta^4(x-y) - e < 0|T((\gamma.A(x)\psi(x))_l\psi_m(y)^*)|0>$$

where

$$A(x) = A_q(x) + A_c(x)$$

$A_q(x)$ denotes the quantum component of the electromagnetic four potential while $A_c(x)$ denotes the purely classical component. This equation can equivalently be expressed in matrix notation as

$$(i\gamma.\partial - m)S(x,y) = \delta^4(x-y) - e < 0|T(\gamma.A(x)\psi(x).\psi(y)^*)|0>$$

The last term on the rhs is called the vertex function. In can be approximated by

$$< 0|T(\gamma.A(x)(\int S_0(x,z)\gamma.A(z)\psi(z)d^4z)\psi(y)^*)|0>$$

$$= \int \gamma^\mu S_0(x,z) < 0|T(A_\mu(x)\gamma^\nu A_\nu(z)\psi(z)\psi(y)^*)|0> d^4z$$

$$= \int \gamma^\mu S_0(x,z)\gamma^\nu S_0(z,y)D_{0\mu\nu}(x,z)d^4z$$

where $S_0(x,y)$ and $D_{0\mu\nu}(x,z)$ are the bare electron and photon propagators. Since the bare propagators are functions only of the difference between the space-time coordinates, we can express the above approximation as

$$\int \gamma^\mu S_0(x-z)\gamma^\nu S_0(z-y)D_{0\mu\nu}(x-z)d^4z$$

$$= \int \gamma^\mu S_0(z)\gamma^\nu S_0(x-y-z)D_{0\mu\nu}(z)d^4z = V(x-y)$$

say. The approximately corrected electron propagator is then given by

$$S(x,y) = S_0(x-y) - e\int S_0(x-z)V(z-y)d^4z$$

or equivalently in kernel notation,

$$S = S_0 - eS_0V$$

The inverse of the electron propagator is approximately

$$S^{-1} = (S_0(1-eV))^{-1} = (1+eV)S_0^{-1} = S_0^{-1} + eVS_0^{-1} = \gamma.p - m_0 + eV(\gamma.p - m_0)$$

We can express this as

$$Z(p)(\gamma.p - m(p))$$

where $m(p)$ is the mass of the electron which equals the sum of its bare value m_0 plus radiative corrections coming form the term V. The computation of the radiative corrections to the electron mass using this method has been discussed in David Atkinson and Peter Wear Johnson. Now if there is an additional random classical component to the electromagnetic field having zero mean and small autocorrelation function, the question is how to evaluate the large deviation rate function for this corrected mass, in particular, how to evaluate using this rate function, the probabilty that the mass of the electron will fall within a certain range of its bare value. The vertex function

$$\int \gamma^\mu S_0(x,z) < 0|T(A_\mu(x)\gamma^\nu A_\nu(z)\psi(z)\psi(y)^*)|0 > d^4z$$

discussed above will involve two components, owing to the decomposition

$$A_\mu = A_{q\mu} + A_{c\mu}$$

Thus,

$$< 0|T(A_\mu(x)A_\nu(z))|0 >= D_{q\mu\nu}(x,z) + D_{c\mu\nu}(x,z)$$

where

$$D_{q\mu\nu}(x,z) =< 0|T(A_{q\mu}(x)A_{q\nu}(z))|0 >, D_{c\mu\nu}(x,z) =< A_{c\mu}(x)A_{c\nu}(z) >$$

D_q is the quantum bare photon propagator and $D_{cmu\nu}$ is the classica photon correlation function. If we assume that D_c is small and that $A_{c\mu}(x)$ is a zero mean Gaussian random field, then we can ask the question of how much will the effective electron mass change due to the classical correction. More precisely, we should use for the photon propagator

$$D_{\mu\nu}(x,z) = D_{q\mu\nu}(x,z) + A_{c\mu}(x)A_{c\nu}(z)$$

and then

$$V(x-y) = \int \gamma^\mu S_0(z)\gamma^\nu S_0(x-y-z)D_{0\mu\nu}(z)d^4z$$

$$+ \int \gamma^\mu S_0(x,z)\gamma^\nu S_0(z,y)A_{c\mu}(x)A_{c\nu}(z)d^4z = V_0(x-y) + \delta V(x,y)$$

The first term is the quantum component of the vertex function and is non-random while the second term is the classical component and is a quadratic form in a classical zero mean Gaussian field. The effect of the random term $\delta V(x,y)$ on the effective mass of the electron can be evaluated by noting that the inverse of the approximate corrected electron propagator is given by

$$\gamma.p - m_0 + eV(\gamma.p - m_0) = \gamma.p - m_0 + e(V + \delta V).(\gamma.p - m_0)$$

In case the bare electron mass m_0 is zero, the electron mass m acquired from radiative corrections is given by the equation

$$Z(p)(\gamma.p - m) = \gamma.p + e(V + \delta V)\gamma.p$$

Chapter 12

How the Electron Acquires Its Mass, Estimating the Electron Spin and the Quantum Electromagnetic Field Within a Cavity in the Presence of Quantum Noise

12.1 Large deviation methods in classical and quantum field theory

[1] Corrections to the electron mass from electro-weak interactions.

The electron-lepton field is ψ_e. Its Lagrangian density after taking into account interactions with the gauge fields $A^a_\mu, a = 1, 2, 3$ and B_μ is given by

$$L_e = \bar{\psi}_e \gamma^\mu (i\partial_\mu + g_1 A^a_\mu \tau_a + g_2 B_\mu)\psi$$

This Lagrangian density can also be expressed in the form

$$L_e = \bar{\psi}_e \gamma^\mu (g_1 A_\mu.\tau + g_2 B_\mu)\psi_e = \bar{\psi}_e \gamma.(g_1 A.\tau + g_2 B)\psi_e$$

The Higgs doublet scalar Lagrangian density is given by

$$L_s = (D_\mu \phi)^* (D^\mu \phi)$$

where D_s is the gauge covariant derivative defined by

$$D_\mu = \partial_\mu - i(g_1 A^a_\mu \tau_a + g_2 B_\mu)$$

We now apply a linear transformation on the gauge fields in such a way that when the Higgs field falls into its ground state value, then the contribution of the Higgs scalar Lagrangian from the gauge fields consists of two decoupled quadratic forms corresponding repsectively to a complex combination of the gauge fields, a quadratic form corresponding to the other real combination of the gauge fields and the final remaining gauge field quadratic form does does not appear. The appearance of such quadratic combinations has the interpretation that there are two gauge bosons which have masses and correspond to the propagators of the nuclear forces while the third gauge boson is massless and is the photon which corresponds to the propagation of the electromagnetic forces.

Define

$$W_\mu = A^1_\mu + iA^2_\mu, W^*_\mu = A^1_\mu - iA^2_\mu,$$

$$Z_\mu = c(\theta)A^3_\mu - s(\theta)B_\mu, D_\mu = s(\theta)A^3_\mu + c(\theta)B_\mu$$

where $c(\theta) = cos(\theta), s(\theta) = sin(\theta)$. Then The term that is quadratic in the gauge fields within the expression $(D_\mu\phi)^*.(D^\mu\phi)$ is given by

$$[(g_1 A^a_\mu \tau_a + g_2 B_\mu)\phi]^*[(g_1 A^{\mu a}\tau_a + g_2 B^\mu)\phi]$$

$$= (g_1^2\phi^*\tau_a\tau_b\phi)A^a_\mu A^{\mu b} + g_2^2\phi^*\phi.B_\mu B^\mu$$

$$+2g_1 g_2\phi^*\tau_a\phi.A^a_\mu B^\mu$$

Note that the gauge fields A^a_μ, B_μ are real fields while the Higgs scalar field doublet ϕ is complex. The matrices τ_a that are the generators of the gauge group are Hermitian. Now by appropriate selection of the Weinberg angle θ, we can ensure that when the above quadratic form in the gauge fields is expressed in terms of the complex gauge field W_μ and the real gauge fields Z_μ and D_μ, for an appropriate choice of the Higgs ground state field ϕ, then the only terms in the gauge field that appear are and $W^*_\mu W^\mu, Z_\mu Z^\mu$ and the $D_\mu D^\mu$ simply does not appear. This means that after symmetry breaking following coupling to the Higgs field, only the W and Z gauge bosons acquire masses while the photon D_μ remains massless. Now we must discuss the mechanism by which the electron acquires mass. This is achieved via the Yukawa coupling involving the electron field ψ_e and the Higgs field ϕ. This coupling term has the general form $H(\psi^*_e \otimes \psi_e \otimes \phi)$ where H is a row vector. ϕ represents a real ground state of the Higgs field. This term can be expressed in the form $M(\phi)\psi^*_e\gamma^0\psi_e$ with $M(\phi)$ dependent upon the coupling vector H as well as on the ground state ϕ of the Higgs field. By suitably adjusting the normalization constant in ϕ or equivalently H, we can make $M(\phi)$ coincide with the observed electron mass.

The large deviation problem: If the ground state of the Higgs field has a small random component, then what will be the rate function of the electron mass ? Hence, using this rate function, estimate the probability that the electron mass will deviate by an amount more than a prescribed threshold ϵ from the observed electron mass.

[2] The role of gravity in giving extra mass to the electron. Dirac's equation in a gravitational field taking into account non-Abelian gauge fields is given by

$$[e_a^\mu(x)\gamma^a(i\partial_\mu + i\Gamma_\mu(x) + g_1 A_\mu^a(x)\tau_a + g_2 B_\mu(x)) - m]\psi(x) = 0$$

where $e_a^\mu(x)$ is a tetrad basis for the metric field and $\Gamma_\mu = (1/2)e_a^\nu e_{b\nu:\mu}[\gamma^a, \gamma^b]$ is the spinor connection of the gravitational field. The Lagrangian density of the Dirac field in this background field comprising gravity, non-Abelian and Abelian photon fields is given by

$$L = \bar{\psi}(x)[e_a^\mu(x)\gamma^a(i\partial_\mu + i\Gamma_\mu(x) + g_1 A_\mu^a(x)\tau_a + g_2 B_\mu(x)) - m]\psi(x)]\sqrt{-g(x)}$$

Note that

$$\sqrt{-g(x)} = det(((e_\mu^a(x)))$$

Let $\psi_l(x)$ denote the canonical position fields. The canonical momentum fields are

$$\pi_l(x) = \partial L/\partial\partial_0\psi(x) = i(\bar{\psi}(x)e_a^0\gamma_a\sqrt{-g})_l$$

and we have the canonical anticommutation relations

$$\{\psi_l(t,r), \pi_m(t,r')\} = i\delta(l,m)\delta^3(r-r')$$

From these anticommutation relations, we derive

$$\{\psi_l(t,r), e_a^0(t,r')e(t,r')(\gamma^a\gamma^0\psi^*)_m(t,r')\} = \delta(l,m)\delta^3(r-r')$$

or equivalently,

$$(\gamma^a\gamma^0)_{sm}e_a^0(t,r')e(t,r')\{\psi_l(t,r), \psi_s(t,r')^*\} = \delta(l,m)\delta^3(r-r')$$

Note that

$$\sqrt{-g(t,r)} = e(t,r) = det((e_\mu^a(t,r)))$$

We write

$$\alpha^a = \gamma^a\gamma^0, a = 0,1,2,3$$

The $\alpha^{a's}$ are Hermitian matrices whose squares are the identity and we can express the above anticommutation relations in the form

$$(\alpha^a)_{sm}(e_a^0 e)(t,r')\{\psi_l(t,r), \psi_s(t,r')^*\} = \delta(l,m)\delta^3(r-r')$$

Let $\beta(t,r)$ denote the inverse of the matrix $\alpha^a(e_a^0.c)(t,r)$. Then this can be expressed as

$$\{\psi_l(t,r), \psi_m(t,r')^*\} = \beta_{lm}(t,r)\delta^3(r-r')$$

What we require now is the exact differential equation for the electron propagator in a background gravitational field. To get this, we first observe that the electron propagator is defined by

$$S_{lm}(x,x') =< 0|T(\psi_l(x)\psi_m(x')^*)|0>, x = (t,r), x' = (t',r')$$

or equivalently in matrix notation,

$$S(x,x') =< 0|T(\psi(x).\psi(x')^*)|0 >= \theta(t-t') < 0|\psi(x)\psi(x')^*|0 > -\theta(t'-t) < 0|\psi(x'$$

Now consider the Dirac operator

$$D_x = e^{\mu})_a(x)\gamma^a i(\partial_\mu + \Gamma_\mu(x)) - m$$

We can decompose D_x into a temporal part and a spatial part

$$D_x = ie_a^0\gamma^a(\partial_0 + \Gamma_0) + ie_a^r\gamma^a(\partial_r + \Gamma_r)$$

where the roman index r is spatial, ie runs over $r = 1, 2, 3$ in contrast with the Greek index μ which is a space time index running over $\mu = 0, 1, 2, 3$. Now,

$$\partial_0 S(x,x') = \delta(t-t') < \{\psi(t,r), \psi(t,r')^*\} > + < T(\partial_0\psi(x).\psi(x')^*) >$$

$$= \beta(t,r)\delta^4(x-x') + < T(\partial_0\psi(x).\psi(x')^*) >$$

Also,

$$\partial_r S(x,x') == < T(\partial_r\psi(x).\psi(x')^*) >$$

and hence

$$D_x S(x,x') = \beta(t,r)\delta^4(x-x')$$

which means that formally, the exact electron propagator is given by

$$S(x,x') = D_x^{-1}(\beta(x)\delta^4(x-x'))$$

In this formalism, we are assuming that the background gravitational field is classical. If however, it is also a quantum field, then vertex functions involving averaged time ordered products of the tetrad field with the Dirac field will also appear in the expression of the exact electron propagator. From the corrected electron propagator, its corrected mass due to interaction with gravity and radiation of electromagnetic and non-Abelian gauge fields can be determined.

[3] The aim of this work is to estimate the electromagnetic field at a given spatial point within the cavity as well as the spin of the electron bound to its nucleus placed within the cavity. The bath surrounding the cavity consists of a noisy electromagnetic field which interacts with both the cavity field and the spin of the electron placed within the cavity. The cavity em field Hamiltonian is given by $H_{01} = \sum_{k=1}^p omega(k)a(k)^*a(k)$ while the electron Hamilonian is given by $H_{02} = \alpha(\sigma, B(t)) + P_\xi^2/2m + V(\xi)$ where $P_\xi = -i\nabla_\xi$ and $B(t)$ is the sum of a classical external magnetic field and the cavity quantum magnetic field expressible as a superposition of the $a(k)'s$ and that $a(k)^{*'s}$. Specifically we can write

$$B(t) = B_0(t) + \sum_{k=1}^p (a(k)f_k(\xi)a_k + \bar{f}_k(\xi)a_k^*)$$

where $B_0(t)$ is the external classical magnetic field and ξ is the electron position relative to its nucleus. The electron observables are $\sigma = (\sigma_x, \sigma_y, \sigma_z)$ and ξ ie its spin and position. These variables form a complete system of observables for the electron just as the $(a(k) + a(k)^*)/2, k = 1, 2, ..., p$ form a complete system of observables for the cavity em field. The system Hilbert space $\mathfrak{h} = \mathfrak{h}_1 \otimes \mathfrak{h}_2$ where $\mathfrak{h}_1$ is $\Gamma_s(\mathbb{C}^p)$ which is isomorphic to $L^2(\mathbb{R}^p)$ while $\mathfrak{h}_2 = L^2(\mathbb{R}^3) \otimes \mathbb{C}^2$. $\mathfrak{h}_1$ is the Hilbert space in which the cavity field operators a_k, a_k^* act while $\mathfrak{h}_2$ is the Hilbert space in which the electron's position, momentum and spin operators act. The bath electromagnetic field interacts with the cavity electromagnetic field with an interaction Hamiltonian (obtained by superposing the cavity field with the bath field, squaring this superposition, integrating it over the cavity volume and considering only cross terms)

$$H_1(t) = \sum_{k,l} (C_1(k,l) a_k dA_l(t)/dt + C_2(k,l) a_k dA_l(t)^*/dt + h.c.)$$

where $h.c.$ denotes Hermitian conjugate of the previous terms. The bath noisy magnetic field also interacts with the electron spin within the cavity with interaction Hamiltonian

$$H_2(t) = \beta(\sigma, \sum_k (c(k) dA_k(t)/dt + \bar{c}(k) dA_k(t)/dt)$$

where $c(k)$ is a complex 3×1 vector that defines the bath noisy magnetic field. Thus the evolution operator $U(t)$ of the cavity system and bath has the form

$$dU(t) = (-i(H_{01} + H_{02} + P)dt - i(C_1(k,l) a_k$$
$$+ C_2(k,l) a_k^* + \beta(\sigma, c(l)) dA_l(t) - i(\bar{C}_1(k,l) a_k^* + \bar{C}_2(k,l) a_k$$
$$+ \beta(\sigma, \bar{c}(l))) dA_l(t)^*) U(t)$$

The term Pdt is the quantum Ito correction term added to ensure unitarity of the joint system-bath evolution. Let $H_0 = H_{01} + H_{02}$ and

$$L_l = -i((C_1(k,l) a_k + C_2(k,l) a_k^* + \beta(\sigma, c(l)), l = 1, 2, ..., q$$

Then, the above HPS-qsde can be expressed in compact notation as

$$dU(t) = (-(iH_0 + P)dt + \sum_{l=1}^{q} (L_l dA_l(t) - L_l^* dA_l(t)^*)) U(t), P = (1/2) \sum_l L_l L_l^*$$

Filtering for estimating the electron spin: Here, we take as our observavble $X = (n, \sigma) = n_x \sigma_x + n_y \sigma_y + n_z \sigma_z$ where $n = (n_x, n_y, n_z)$ is a real unit 3-vector. After time t, this evolves to

$$j_t(X) = U(t)^* X U(t)$$

and it satisfies the Evans-Hudson flow equation

$$dj_t(X) = j_t(\theta_0(X))dt + j_t(\theta_{1l}(X)) dA_l(t) + j_t(\theta_{2l}(X)) dA_l^*(t)$$

where

$$\theta_{1l}(X) = [X, L_l], \theta_{2l}(X) = [L_l^*, X], \theta_0(X) = -PX - XP + \sum_l L_l X L_l^*$$

The input measurement process is

$$Y_i(t) = \sum_l *d(l)A_l(t) + \bar{d}(l)A_l(t)^*)$$

and the output measurement is then

$$Y_o(t) = U(t)^* Y_i(t) U(t)$$

Thus,

$$dY_o(t) = dY_i(t) - j_t(\sum_l \bar{d}(l)L_l + d(l)L_l^*)dt$$

Thus, this measurement model amounts to measuring a quantum white Brownian noise corrupted version of the observable $Z = -\sum_l d(l)L_l^* + \bar{d}(l)L_l$. Note that

$$i\sum \bar{d}(l)L_l = \sum_l \bar{d}(l)((C_1(k,l)a_k + C_2(k,l)a_k^* + \beta(\sigma, c(l))$$

So if we choose that $d(l)'s$ in such a way that

$$\sum \bar{d}(l)C_1(k,l) = \sum_l \bar{d}(l)C_2(k,l) = 0\forall k$$

then essentially, this measurement model amounts to measuring a noise corrupted linear combination of the Pauli spin matrices with noise. Likewise, if we choose the $d(l)'s$ so that

$$\sum_l \bar{d}(l)c(l) = 0$$

then this measurement model amount to measuring a noise corrupted linear combination of the $a_k's$ and a_k^*'s, which is equivalent to measuring a component of the quantum electromagnetic field within the cavity at a given spatial point. Note that we cannot measure the field at more than one spatial point in general simultaneously because the field at two different spatial points need not commute.

Some remarks on the measurement process: We can generally measure two kinds of non-demoltion processes, the first is quantum Brownian motion which is a linear combination of the creation and annihilation processes, the second is quantum Poisson noise which is a linear combination of the conservation processes. Measuring the former amounts to measuring the quantum electromagnetic field since the electromagnetic vector potential within the cavity is a linear combination of the cavity field creation and annihilation operators and the Lindblad operators that modulate the creation and annihilation processes are also linear combination of the cavity field creation and annihilation operators. Measuring the latter means that our output measurement process is

$$Y_o(t) = U(t)^* Y_i(t) U(t), Y_i(t) = \sum_l e(l)\Lambda_l(t), d\Lambda_l(t) = dA_l(t)^* dA_l(t)/dt$$

Then, we find that

$$dY_o(t) = dY_i(t) + j_t(F_l)dA_l + j_t(G_l)dA_l^*$$

where F_l, G_l are again linear combinations of the L_l, L_l^* which are linear combinations of the cavity field creation and annihilation operator. On the other hand, suppose we include in the HPS-QSDE conservation operator terms of the form $(S_l - 1)d\Lambda_l(t)$ where $S_l^* S_l = 1$ to guarantee unitary evoluition. Then in our Quantum Poisson measurement model, we get extra terms $j_t(S_l^* + S_l - 2)d\Lambda_l(t)$. Now what is the physical nature of the coefficients S_l of the quantum Poisson/conservation processes in the HP-QSDE ?. if we have N_B bath photons interacting with N_S cavity photons, then their interaction energy will be proportional to the product $N_B N_S$ and hence S_l should be linear combinations of the cavity modal photon numbers $a_k^* a_k$. Thus, using a quantum Poisson process measurement model amounts to measuring the intensity of photons in the cavity. Another way to see this is to consider such photon number/intensity interactions as being mediated by electrons or some other elementary particles. When bath photons hit a photocell, they generates an electronic current proportional to the number of photons that have frequencies/energies more than the work function of the electrons. Likewise when cavity photons hit a photocell, they generate electronic current proportional to the number of cavity phtons. The interaction between these two currents takes place because of the interaction between the electromagnetic field generated by these two currents and this interaction energy is therefore proportional to the product of the two currents which is in turn proportional to the product of the number of bath and cavity photons.

Our aim is to simulate the evolution of a system state observable X which after time t, evolves to $j_t(X) = U(t)^* X U(t)$ and along with it to simulate its estimate $\pi_t(X) = \mathbb{E}(j_t(X)|\eta_o(t))$ based on the non-demolition Abelian output algebra $\eta_o(t) = \sigma(Y_o(s) : s \leq t)$ and to plot the error process

$$E_t(X) = j_t(X) - \pi_t(X)$$

which will be a family of Hermitian matrices which operate on the tensor product of the system and bath Hilbert spaces. The nsr (noise to signal ratio)

$$n(t) = \frac{\| E_t(X) \|}{\| j_t(X) \|} = \frac{\| E_t(X) \|}{\| X \|}$$

will also be plotted for different kinds of system observables X like the components of the Pauli spin matrices and also linear combinations of the cavity field creation and annihilation operators. Our simulations will demonstrate that these nsr's will eventually become much smaller than unity validating thereby the effectiveness of the Belavkin quantum filter. In order to simulate the Belavkin filter, we asssume that the bath noise processes consist of just a single creation and a single annihilation process. For dealing with this case, it suffices to let our bath Hilbert space to be the Boson Fock space $\Gamma_s(L^2[0,T])$. we choose a large set of normalized linearly independent vectors $u_1,...,u_N$ in $L^2([0,T])$ such as sinusoids $u_n(t)=\sqrt{2/T}.sin(2\pi nt/T), n=1,2,...,N$ and then Gram-Schmidt orthonormalize the corresponding exponential vectors $e(u_n), n=1,2,...,N$ in $L^2([0,T])$ using the well known relation (K.R.Parthasarathy, "An introduction to quantum stochastic calculus)

$$<e(u),e(v)>=exp(<u,v>)]$$

which therefore imply

$$e(u_n),e(u_m)>=exp(\delta[n-m])=1ifn\neq m,e,ifn=m$$

We denote these Gram-Schmidt orthonormalized vectors by $e_n, n=1,2,...,N$ so we can write

$$e(u_n)=\sum_{k=1}^{N}c(n,k)e_k, e_n=\sum_{k=1}^{N}d(n,k)e(u_k)$$

with the coefficients being such that $c(n,k)=d(n,k)=0$ for $k>n$. Now choose an orthonormal basis $\{f_1,...f_r\}$ for the system Hilbert space $\mathfrak{h}$ and construct an approximate orthonormal basis $\{f_m\otimes e_s : 1\leq m\leq r, 1\leq s\leq N\}$ for the tensor product $\mathfrak{h}\otimes L^2([0,T])$ of the system Hilbert space and the bath Hilbert space. We then discretize time into steps of width δ. As an example of how we carry out our simulations, we consider the qsde

$$dU(t)=(L_0dt+L_1dA(t)+L_2dA(t)^*)U(t)$$

We approximate $U(t)$ by an $rN\times rN$ matrix and then replace the above evolution equation by

$$d<f_m\otimes e_s|U(t)|f_n\otimes e_l>=\sum_{q,k}<f_m|L_0|f_q>\delta(s,k)<f_q\otimes e_k|U(t)|f_n\otimes e_l>dt$$

$$+\sum_{q,k}<f_m|L_1|f_q><e_s|dA(t)|e_k><f_q\otimes e_k|U(t)|f_n\otimes e_l>$$

$$+\sum_{q,k}<f_m|L_1|f_q><e_s|dA(t)^*|e_k><f_q\otimes e_k|U(t)|f_n\otimes e_l>$$

where

$$<e_s|dA(t)|e_k>=\sum_{m,n}\bar{d}(s,m)d(k,n)<e(u_m)|dA(t)|e(u_n)>,$$

$$<e(u_m)|dA(t)|e(u_n)>=u_n(t)<u_m|u_n>dt=u_n(t)\delta(m,n)dt$$

and likewise

$$<e_s|dA(t)^*|e_k>=\sum_{m,n}\bar{d}(s,m)d(k,n)<e(u_m)|dA(t)^*|e(u_n)>,$$

$$<e(u_m)|dA(t)^*|e(u_n)>=\bar{u}_m(t)<u_m|u_n>dt=\bar{u}_m(t)\delta(m,n)dt$$

Chapter 13

Mathematical Tools for Large Deviations, Neural Networks, LDP in Physical Theories, EM and LDP Algorithms in Quantum Parameter Estimation and Filtering

Mathematical tools for large deviation theory

13.1 Lecture on the Ascoli Arzela theorem and Prohorov's tightness theorem with applications to proving weak convergence of probability distributions on the space of continuous functions on a compact interval

Let (X, d) be a compact metric space and let $C(X)$ denote the space of all real valued continuous functions on X. Introduce a metric ρ on $C(X)$ by the formula

$$\rho(f, g) = sup_{x \in X}|f(x) - g(x)|$$

$(C(X), \rho)$ is a complete metric space. Also any $f \in C(X)$ is uniformly continuous. To prove these facts, let $f_n \in C(X)$ be Cauchy, ie, $\rho(f_n, f_m) \to 0$ as $n, m \to \infty$. Then, for each $x \in X$ we have $|f_n(x) - f_m(x)| \to 0$ and by completeness of $\mathbb{R}$, it follows that for each $x \in X$, there is an $f(x) \in \mathbb{R}$ such that $f_n(x) \to f(x)$. It is clear that f_n converges to f in the uniform metric ρ since

$$|f_n(x) - f_x)| \leq \rho(f_n, f_m) + |f_m(x) - f(x)|$$

and hence

$$|f_n(x) - f(x)| \leq limsup_m \rho(f_n, f_m), \forall x \in X$$

so

$$\rho(f_n, f) \leq limsup_m \rho(f_n, f_m) \to 0, n \to \infty$$

Further, it is clear that f is continuous since

$$|f(x)-f(y)| \leq |f(x)-f_n(x)|+|f_n(x)-f_n(y)| \\ +|f_n(y)-f(y)| \leq \rho(f, f_n)+|f_n(x)-f_n(y)|+\rho(f_n, f)$$

Given any $\epsilon > 0$, choose n so that $\rho(f, f_n) < \epsilon$ and then choose $\delta > 0$ so that for a given x, $d(y, x) < \delta$ implies $|f_n(x) - f_n(y)| < \epsilon$. This is possible because f_n is continuous. Then, it follows that $d(y, x) < \delta$ implies

$$|f(x) - f(y)| < 3\epsilon$$

proving that f is continuous. Thus, we have proved that $(C(X), \rho)$ is a complete metric space. Now we prove that any $f \in C(X)$ is also uniformly continuous. Let $\epsilon > 0$ be given. For each $x \in X$, choose a $\delta(x) > 0$ such that $y \in B(x, \delta(x))$ implies $|f(y) - f(x)| < \epsilon$. Since $X = \bigcup_{x \in X} B(x, \delta(x)/3)$, by compactness of X, there exists a finite set $\{x_1, ..., x_N\} \subset X$ such that

$$X = \bigcup_{k=1}^{N} B(x_k, \delta(x_k)/3)$$

Let $\delta = min(\delta(x_k)/3 : k = 1, 2, ..., N)$

Since N is a finite integer, $\delta > 0$ and further, let $x, y \in X, d(x, y) < \delta$. Then, $x \in B(x_k, \delta(x_k)/3)$ for some $k = 1, 2, ..., N$. Then, $d(y, x_k) \leq d(y, x)+d(x, x_k) \leq 2\delta(x_k)/3$ and hence

$$|f(x) - f(x_k)| < \epsilon, |f(y) - f(x_k)| < \epsilon$$

Thus,

$$|f(x) - f(y)| < 2\epsilon$$

and this proves the uniform continuity of f.

13.2 Proof of the Prohorov tightness theorem

Let X be a separable metric space. By the Urysohn theorem, X can be embedded inside a compact metric space Y, such that X is dense in Y. Formally, this means that there exists a compact metric space Y and a map $\phi : X \to Y$ such that ϕ is injective, continuous and $\phi^{-1} : \phi(X) \to X$ is also continuous and finally that $\phi(X)$ is dense in Y. Now, since Y is compact, $C(Y)$ is a separable metric space in view of the Stone-Weierstrass theorem. Specifically, Y being compact, ia also separable and hence there exists a countable dense subset $\{x_n : n = 1, 2, ...\}$ of Y. Then the set of functions $f_n(x) = d(x_n, x), n \geq 1$ separates the points of Y and hence the algebra $\mathcal{A}$ generated by these functions is dense in $C(Y)$ by the Stone-Weierstrass theorem. Now the set of all functions of the form $1, \sum_{n_1,...,n_K=1}^{N} c(n_1, ..., n_K) f_{n_1} ... f_{n_K}$ with $N, K = 1, 2, ...$ and $c(n_1, ..., n_K)$ rational is countable and also dense in $\mathcal{A}$ anbd this proves that $C(Y)$ is a separable metric spacc. Without loss of generality, we can assume that X is a dense subset of Y by renaming $\phi(X)$ as X.

Now choose a countable dense subset $\{g_n : n = 1, 2, ...\}$ in $C(Y)$ an let f_n be the restriction of g_n to X. Let M be tight family of probability measures on X, ie, for every $\epsilon > 0$ there exists a compact set $K_\epsilon \subset X$ such that $\mu(K_\epsilon) > 1 - \epsilon) forall \mu \in M$. Then we wish to show that M is compact, ie, given any infinite sequence $\mu_n, n = 1, 2, ...$ in M, there exists a subsequence μ_{n_m} that converges weakly to some probability measure μ on X. For any measure μ on X, we define the measure $\hat{\mu}$ on Y by $\hat{\mu}(B) = \mu(X \cap B), B \in \mathcal{B}(X)$. It is clear that if μ is a probability measure on X, the $\hat{\mu}$ is also a probability measure on Y. Now for any probabiliy measure ν on Y, define

$$T(\nu) = \{\int g_n d\nu : n \geq 1\}$$

It is clear that $T(\nu)$ is a sequence in $[0, 1]$ provided that we replace g_n by $g_n / \| g_n \|$. Further if ν and μ are two distinct probability measures on Y, then $T(\nu) \neq T(\mu)$. For suppose $T(\nu) = T(\mu)$. Then

$$\int g_n d\nu = \int g_n d\mu \forall n$$

Let g be any continous function on Y and let $\epsilon > 0$. Then there exists an n and a constant $c(n)$ such that

$$\| c(n) g_n - g \| < \epsilon$$

It follows then by the triangle inequality for complex numbers that

$$|\int g d\nu - \int g d\mu| \leq \int |g - c(n) g_n| d\nu$$

$$+ \int |g - c(n) g_n| d\mu + |c(n)| |\int g_n d\nu - \int g_n d\mu| < 2\epsilon$$

and since $\epsilon > 0$ is arbitrary, we get

$$\int g d\nu = \int g d\mu$$

Since g is an arbitrary continuous function on Y, it follows that $\nu = \mu$. In fact, let C be any closed subset of Y. Then we define $G_n = \{y : d(y, C) < 1/n\}$. G_n is a sequence of open sets that decreases to C and we can choose a continous function h_n on Y such that $h_n|_C = 1, h_n|_{G_n^c} = 0$ and $0 \leq h_n \leq 1$. Then

$$\nu(C) \leq \int_C h_n d\nu = \int_C h_n d\mu \leq \mu(G_n)$$

and letting $n \to \infty$ gives us

$$\nu(C) \leq \mu(C)$$

Interchanging the ν and μ in this argument then gives us the result that ν and μ coincide on all the closed subsets of Y. Now define $\mathcal{C}$ to be the collection of all Borel subsets B of Y for which $\nu(B) = \mu(B)$. Then, $\mathcal{C}$ is a σ-field containing all the closed subsets of Y. Thus, $\mathcal{C}$ must coincide with $\mathcal{B}(X)$ and the proof of the claim, namely that T is an injective map is complete.

13.3 Some remarks on neural networks related to large deviation theory

[1] Given a firing nn with p delays for the firing component, the weight update equations can be expressed in the form

$$w(n+1) = F(w(n), w(n-1), ..., w(n-p), \mathbf{x}(n), \mathbf{d}(n))$$

where now $\mathbf{x}(n)$ and $\mathbf{d}(n)$ are to be regarded as vector input and output signals comprising of p delayed versions of the actual input and output. More specifically, if $x_0(n), d_0(n)$ are the true input and output signals which will generally be $\mathbb{R}^L$ vector valued processes, then

$$x(n) = [x_0(n)^T, x_0(n-1)^T, ..., x_0(n-p)^T]^T \in \mathbb{R}^{L(p+1)}, d(n) = [d_0(n)^T, d_0(n-1)^T, ..., d_0(n-p)^T]^T \in \mathbb{R}^{L(p+1)}$$

Now linearizing the above weight update equation around nominal input, output and weight processes, taking into account in addition, small random uncertainties in the measurement of the input and output processes for supply to the nn, we obtain the linearized stochastic difference equation

$$\delta w[n+1] = F_1(\mathbf{w}(n), \mathbf{x}(n), \mathbf{d}(n))\delta\mathbf{w}(n) + F_2(\mathbf{w}(n), \mathbf{x}(n), \mathbf{d}(n))\delta\mathbf{x}(n) + F_3(\mathbf{w}(n), \mathbf{x}(n), \mathbf{d}(n))\delta\mathbf{d}(n)$$

where

$$\mathbf{w}(n) = [w(n)^T, w(n-1)^T, ..., w(n-p)^T]^T$$

and this equation can in turn be expressed in vector form using the formula

$$\delta\mathbf{w}(n+1) = [\delta w(n+1)^T, \delta w(n)^T, ..., \delta w(n-p)^T]^T$$

as

$$\delta\mathbf{w}(n+1) = \mathbf{F}_1(\mathbf{w}(n), \mathbf{x}(n), \mathbf{d}(n))\delta\mathbf{w}(n) + \mathbf{F_2}(\mathbf{w}(n), \mathbf{x}(n), \mathbf{d}(n))\delta\mathbf{x}(n) + \mathbf{F_3}(\mathbf{w}(n), \mathbf{x}(n), \mathbf{d}(n))\delta\mathbf{d}(n)$$

or in shorthand notation, as

$$\delta\mathbf{w}(n+1) = \mathbf{F}_1(n)\delta\mathbf{w}(n) + \mathbf{F_2}(n)\delta\mathbf{x}(n) + \mathbf{F_3}(n)\delta\mathbf{d}(n)$$

Now assume that the input and output uncertainties $\delta\mathbf{x}(n), \delta\mathbf{d}(n), n \geq 1$ are iid processes with lograrithmic moment generating functions

$$\Lambda_x(\lambda) = log\mathbb{E}exp(\lambda^T\delta\mathbf{x}(n)), \Lambda_d(\lambda) = log\mathbb{E}exp(\lambda^T\delta\mathbf{d}(n))$$

Lets calculate the logarithmic moment generating functional of the process $\delta\mathbf{w}(n), 1 \leq n \leq N$:

$$M_w(\lambda(1), ..., \lambda(N)) = \mathbb{E}[exp(\sum_{k=1}^{N}\lambda(k)^T\delta\mathbf{w}(k))] = logM_w(\lambda(1), ..., \lambda(N))$$
$$= \Lambda_w(\lambda(1), ..., \lambda(N))$$

where

$$\delta\mathbf{x}(n) = \sum_{k=1}^{n}\Phi(n,k)(F_2(k)\delta x(k) + F_3(k)\delta d(k)), n \geq 1$$

where $\Phi(n,k), n \geq k$ is the state transition matrix associated with the forcing vector $\mathbf{F}_1(n)$. More specifically, it satisfies

$$\Phi(n+1,k) = \mathbf{F_1}(n)\Phi(n,k), n \geq k, \Phi(n,n) = I$$

Then,

$$\sum_{n=1}^{N}\lambda(n)^T\delta\mathbf{x}(n)$$

$$= \sum_{1\leq k\leq n\leq N}\lambda(n)^T\Phi(n,k)F_2(k)\delta x(k) + \sum_{1\leq k\leq n\leq N}\lambda(n)^T\Phi(n,k)F_3(k)\delta d(k)$$

and therefore

$$\Lambda_w(\lambda(1), ..., \lambda(N)) = log\mathbb{E}[exp(\sum_{n=1}^{N}\lambda(n)^T\delta\mathbf{x}(n))] =$$

$$\sum_{k=1}^{N}\Lambda_x(\sum_{n=k}^{N}F_2(k)^T\Phi(n,k)^T\lambda(n))$$
$$+\sum_{k=1}^{N}\Lambda_d(\sum_{n=k}^{N}F_3(k)^T\Phi(n,k)^T\lambda(n))$$

Now replace the noise processes $\delta x(n)$ and $\delta d(n)$ by $\sqrt{\epsilon}.\delta x(n)$ and $\sqrt{\epsilon}.\delta d(n)$. Then, the limiting scaled logarithmic moment generating function of the weight perturbation process over the time duration $[0, N]$ is given by

$$\bar{\Lambda}_w(\lambda(1), ..., \lambda(N)) = lim_{\epsilon\to 0}a(\epsilon).\Lambda_w(a(\epsilon)^{-1}\epsilon^{1/2}\lambda(1), ..., a(\epsilon)^{-1}\epsilon^{1/2}\lambda(N)) =$$

$$\sum_{k=1}^{N}[\bar{\Lambda}_x(\sum_{n=k}^{N}F_2(k)^T\phi(n,k)\lambda(n)) + \bar{\Lambda}_d(\sum_{n=k}^{N}F_3(k)^T\Phi(n,k)\lambda(n))]$$

where

$$\bar{\Lambda}_x(\lambda) = lim_{\epsilon\to 0}a(\epsilon).\Lambda_x(a(\epsilon)^{-1}\epsilon^{1/2}\lambda^T\delta x(n))$$
$$\bar{\Lambda}_d(\lambda) = lim_{\epsilon\to 0}a(\epsilon).\Lambda_x(a(\epsilon)^{-1}\epsilon^{1/2}\lambda^T\delta d(n))$$

13.4 LDP problems in general relativity and non-Abelian gauge field theory

mu [1] Let $A = A^a_\mu T_a dx^\mu$ denote the YM gauge potential field where the $T^{a's}$ are the generators of the Lie algebra of the gauge group. The YM field tensor is

$$F = dA + g[A, A] = A^a_{\mu,\nu}T_a dx^\nu \wedge dx^\mu + gA^a_\mu A^b_\nu[T_a, T_b]dx^\mu \wedge dx^\nu$$

Let $C(abc)$ denote the structure constants of the gauge group corresponding to the Lie algebra generators $\{T_a\}$. Thus,

$$[T_a, T_b] = C(abc)T_c$$

summation over the index c being implied. Then, the above equation can be expressed as

$$F = A^a_{\mu,\nu}T_a dx^\nu \wedge dx^\mu + gC(abc)A^a_\mu A^b_\nu T_c dx^\mu \wedge dx^\nu = F^a T_a$$

where

$$F^a = A^a_{\mu,\nu}dx^\nu \wedge dx^\mu + gC(bca)A^b_\mu A^c_\nu dx dx^\mu \wedge dx^\nu$$
$$= [(1/2)(A^a_{\nu,\mu} - A^a_{\mu,\nu}) + gC(bca)A^b_\mu A^c_\nu]dx^\mu \wedge dx^\nu = (1/2)F^a_{\mu\nu}dx^\mu \wedge dx^\nu$$

where

$$F^a_{\mu\nu} = A^a_{\nu,\mu} - A^a_{\mu,\nu} + 2gC(bca)A^b_\mu A^c_\nu$$

give explicitly the components of the YM antisymmetric field tensor. Consider now a gauge transformation of the YM potentials:

$$\delta A = d\Lambda + g[A, \Lambda], \Lambda = \Lambda^a T_a$$

where Λ^a are ordinary scalar functions of the space-time coordinates. Under such an infinitesimal gauge transformation, the gauge field F changes by

$$\delta F = \delta(dA + g[A, A]) = d\delta A + g[\delta A, A] + g[A, \delta A]$$
$$= g(d[A, \Lambda] + [d\Lambda, A] + g[[A, \Lambda], A] + [A, d\Lambda] + g[A, [A, \Lambda]])$$
$$= g([dA, \Lambda] + g([[A, \Lambda], A] + [A, [A, \Lambda]])) = g[dA + g.A \wedge A, \Lambda]$$
$$= g[F, \Lambda]$$

This is the coordinate free version of the local Lie group transformation formula

$$F(x) \to g(x).F(x).g(x)^{-1}$$

13.5 Large deviations in String theoretic corrections to field theories

Let ϕ_m denote the field corresponding to the massive string modes and ϕ_0 the field corresponding to the massless modes. The total action of these two fields is $S(\phi_0, \phi_m)$. The effective action for ϕ_0 which describes low energy string theory and therefore represents a string theoretic correction of field theory is given by

$$S_0(\phi_0) = -i.log(\int exp(iS(\phi_0, \phi_m))D\phi_m)$$

Typically, if $\phi(x)$ represents the field of our theory with action functional

$\int L(\phi(x), \phi_{,\mu}(x))d^D x$, its string theoretic correction will be of the form

$$\int L(\phi(x + \sum_k a(k)f_k(\sigma)), \phi_{,\mu}(x + \sum_k a(k)f_k(\sigma))d^D x d\sigma$$

where σ is the string length parameter which varies over $[0, 2\pi)$. The string curve is described by the equation

$$X(\sigma) = x + \sum_k a(k)f_k(\sigma)$$

where x is the centre of the string which represents the position of the Tachyon and the $a(k)'s$ are the Fourier series modal coefficients that describe the oscillation modes of the string. The effective field theoretic action is then

$$S_0(\phi) = \int exp(i \int L(\phi(x+\sum_k a(k)f_k(\sigma)), \phi_{,\mu}(x+\sum_k a(k)f_k(\sigma))d^D x d\sigma)\Pi da(k)$$

Note that the $f_k(\sigma)'s$ are complex valued functions of the string parameter length usually of the form $exp(ik\sigma)$. To obtain approximate expressions for $S_0(\phi)$, we write

$$\delta x(\sigma) = \sum_k a(k)f_k(\sigma)$$

and

$$\phi(x + \delta x(\sigma)) = \phi(x) + \sum_{m \geq 1} (D^m \phi(x))\delta x(\sigma)^{\otimes m}/m!$$

$$\partial_\mu \phi(x + \delta x(\sigma)) = \partial_\mu \phi(x) + \sum_{m \geq 1} (D^m \partial_\mu \phi(x))\delta x(\sigma)^{\otimes m}/m!$$

In short, this can be expressed as

$$\phi(x + \delta x(\sigma)) = \phi(x) + \psi(x, \delta x(\sigma)),$$

$$\partial_\mu \phi(x + \delta x(\sigma)) = \partial_\mu \phi(x) + \partial_\mu \psi(x + \delta x(\sigma))$$

Chapter 14

Quantum Transmission Lines, Engineering Applications of Stochastic Processes

14.1 Introduction

In this paper, we have first set up the basic partial differential equations for the line voltage and line current for a transmission line that may be lossy in terms of the distributed parameters, namely, the resistance, inductance, conductance and capacitance of the line per unit length. By representing the line voltage and current as a Fourier series in the spatial length variable of the line, we have transformed these coupled linear first order pde's into an infinite sequence of ode's for the Fourier series components of the line current and voltage. We have then explained how in the lossless case, these differential equations are exactly the same as those that describe the dynamics of an infinite sequence of decoupled harmonic oscillators and which can therefore be derived either from Lagrangian or after applying the Legendre transformation, from a Hamiltonian. Thus, by quantizing this Hamiltonian for an infinite sequence of harmonic oscillators, the equations of motion for the line voltage and current at the quantum scale, can be derived from the Heisenberg matrix mechanics as is conventionally done in quantum mechanics when the classical Hamiltonian equations hold good but with the Poisson bracket replaced by the Lie bracket.

After, this we have explained how in the lossy case (ie when resistance and conductance per unit length are included), the line equations can be derived from a Hamiltonian plus Lindblad terms using the Heisenberg matrix mechanics. In other words, the line equations for the Fourier series components of the line voltage and current describe simply a sequence of infinite damped harmonic oscillators which can be obtained from the Heisenberg-Hamiltonian-Lindblad matrix mechanics for open quantum system. In fact, it is well known that the equations of a damped harmonic oscillator cannot be derived from a a purely Hamiltonian approach because the former is a dissipative system while the Hamiltonian approach works only for conservative systems. Thus we have to add additional Lindblad terms to the Hamiltonian component of the matrix mechanics equations in order to derive damped harmonic oscillator equations and when one adopts this approach for an infinite sequence of damped harmonic oscillators with the natural frequencies appropriately chosen, then we are able to model a lossy transmission line at the quantum scale. This is precisely our approach here. We next consider the dual of this Lindblad equation namely the GKSL equation for the mixed state of the transmission line. Its evolution is the Schrodinger equation for the mixed state plus Lindblad terms and we outline a procedure to solve it by expanding the state using the non-orthogonal Glauber-Sudarshan representation as a superposition of mixed coherent states. The coherent states are associated with the sequence of creation and annihilation operators of the infinite sequence of damped harmonic oscillators appearing in the Hamiltonian and Lindblad operator terms of the quantized transmission line. The GKSL equation is then shown to reduce to a partial differential equation for a complex function of time and the coherent complex variables. Finally, we make a computation on the rate of change of the line entropy caused by the Lindblad dissipative terms. The idea is to start with Von-Neumann's expression for the quantum entropy of the line and evaluate its rate of change from the GKSL equation for the density keeping in mind that while differentiating the logarithm of the density, we make use of the standard formula in Lie algebras regarding the computation of the differential of the exponential map. This enables us to assess the situation of when the line entropy will increase in accordance with the second law of thermodynamics.

14.2 Kolmogorov's existence for stochastic processes applied to the problem of describing infinite image fields, ie, image fields with a countably infinite number of pixels

14.3 Dirichlet series with image processing applications

Consider an image field

$$F(r,s) = \sum_{n,m\geq 1} \chi_1(n)\chi_2(m)/n^r m^s$$

Let P denote the set of primes and consider the Euler-Dirichlet product

$$G(r,s) = [\Pi_{p,q\in P}(1-\chi_1(p)/p^r)(1-\chi_2(p)/q^s)]^{-1} =$$

$$\Pi_{p,q\in P}\sum_{n,m\geq 0}\chi_1(p^n)\chi_2(q^m)/p^{nr}q^{ms}$$

where $\chi_k, k = 1,2$ are multiplicative functions, ie, $\chi_k(nm) = \chi_k(n)\chi_k(m)$ for $k = 1,2$ whenever $gcd(n,m) = 1$. Expanding further,

$$G(r,s) = \sum_{n_1,...,m_1,...,\geq 0}\chi_1(p_1^{n_1}p_2^{n_2}...)\chi_2(q_1^{m_1}q_2^{m_2}...)/[(p_1^{n_1}p_2^{n_2}...)^r(q_1^{m_1}q_2^{m_2}...)^s]$$

$$= \sum_{n,m\geq 1}\chi_1(n)\chi_2(m)/n^rm^s = (\sum_{n\geq 1}\chi_1(n)/n^r).(\sum_{n\geq 1}\chi_2(n)/n^s)$$

$$= F(r,s)$$

More generally, for an arbitrary function $\chi(n,m)$ on $\mathbb{Z}_+^2$, define

$$F(r,s) = \sum_{n,m\geq 1}\chi(n,m)/n^rm^s$$

$$= \sum_{n,m\geq 1}\chi(n,m)\int_{\mathbb{R}_+^2}exp(-nt_1-mt_2)t_1^{r-1}t_2^{s-1}dt_1dt_2/\Gamma(r)\Gamma(s)$$

Writing

$$\psi(t_1,t_2) = \sum_{n,m\geq 1}\chi(n,m)exp(-nt_1-mt_2)$$

we have

$$F(r,s) = \int_{\mathbb{R}_+^2}\psi(t_1,t_2)t_1^{r-1}t_2^{s-1}dt_1dt_2/\Gamma(r)\Gamma(s)$$

In other words, $F(r,s)$ is the Laplace-Mellin transform of $\psi(t_1,t_2)$. Let $\psi(t_1,t_2)$ denote an image field and its scaled version is

$$\psi_{a,b}(t_1,t_2) = \psi(a^{-1}t_1,b^{-1}t_2), a,b>0$$

Let $F_{ab}(r,s)$ be the above Laplace-Mellin transform of ψ_{ab}. Then,

$$F_{ab}(r,s) = \int_{\mathbb{R}_+^2}\psi(a^{-1}t_1,b^{-1}t_2)t_1^{r-1}t_2^{s-1}dt_1dt_2/\Gamma(r)\Gamma(s)$$

$$= a^rb^sF(r,s)$$

This shows that the scale factors a,b of the image field can be estimated from the original image ψ and its scaled version ψ_{ab} in terms of their Laplace-Mellin transforms F, F_{ab} in the presence of noise, ie, when

$$\psi_{ab}(t_1,t_2) = \psi(a^{-1}t_1,b^{-1}t_2)+w(t_1,t_2)$$

using the formula

$$F_{ab}(r,s) = a^r b^s F(r,s) + F_w(r,s)$$

Writing $F_{ab} = G$, we get

$$G(r,s) = a^r b^s F(r,s) + F_w(r,s)$$

where

$$F_w(r,s) = \int_{\mathbb{R}_+^2} w(t_1,t_2) t_1^{r-1} t_2^{s-1} dt_1 dt_2 / \Gamma(r)\Gamma(s)$$

is the Laplace-Mellin transform of the noise $w(t_1,t_2)$. a, b may be estimated by applying the least squares method to

$$ln(G(r,s)) = ln(a^r b^s F(r,s) + F_w(r,s)) \approx r.ln(a) + s.ln(b) + ln(F(r,s))$$

so that writing $ln(a) = \theta, ln(b) = \phi$, their estimates will be given by

$$(\hat{\theta}, \hat{\phi}) = argmin_{(\theta,\phi)} \int (ln(G(r,s)/F(r,s)) - r.\theta - s.\phi)^2 d\mu(r,s)$$

where μ is an appropriate measure on the scaling space $\mathbb{R}_+^2$. The measure μ is obtained from the noise statistics. Specifically, if the noise is of small amplitude, then we have upto linear orders in it the following approximation to the above noisy model in the Laplace-Mellin domain:

$$ln(G(r,s))$$

$$= ln(a^r b^s F(r,s)(1+F_w(r,s)/a^r b^s F(r,s))) \approx ln(a^r b^s F(r,s)) + F_w(r,s)/a^r b^s F(r,s)$$

or equivalently,

$$X(r,s) = ln(G(r,s)/F(r,s)) \approx r.\theta + s.\phi + (F_w(r,s)/F(r,s)).exp(-r\theta - s\phi)$$

and knowing the statistics of $(F_w(r,s)/F(r,s)).exp(-r\theta - s\phi)$ using the statistics of $w(t_1,t_2)$ and approximate values of θ, ϕ, the ML estimate of θ, ϕ can be calculated.

Ref: This problem was suggested to me by my colleague Dr.Neeraj.

14.4 About the book

This book talks about non-linear and quantum mechanical phenomena in transmission lines and waveguides as well as about statistical and quantum statistical methods for estimating the parameters of transmission lines and waveguides from measurements of the transition probabilities of an electron bound to its nucleus when a classical and a quantum electromagnetic field in space generated by the line current or within a waveguide falls on the atom with the field being in a coherent state. While discussing nonlinear phenomena in transmission lines, using the basic principles of physics we derive general nonlinear relationships between the magnetic field and magnetization and between the electric field and polarization in the form of nonlinear systems having memory and from these relationships, we establish general nonlinear relationships between the line voltage and the line current and then analyze the solutions to these nonlinear integro-differential equations using perturbation theory for nonlinear systems. Regarding quantum aspects of waveguides and transmission lines, we also make some computations about the rate of change of the Von-Neumann entropy of the atomic state when quantum electromagnetic radiation falls upon it with the radiation being in a coherent state. These computations find applications to communication problems where we wish to assess the information received by an atomic receiver from the quantum electromagnetic field arriving from the transmitter antenna on the nano-scale. The book also contains some material on large deviation theory and its application to engineering systems, a field in applied probability which recently has gained importance because of its ability to predict probabilities of rare spikes of noise in a dynamical system so that such spikes can be controlled to reduce the chance of damage of the system.

14.5 An application of the EM algorithm to quantum parameter estimation and quantum filtering

Consider an electron bound to its nucleus. Let N be the number of protons in the nucleus and let e and m denote respectively the charge and mass of the electron. The parameter vector to be estimated is $\theta = (N, m, e)$. The unperturbed electronic Hamiltonian is

$$H_0 = -\nabla^2/2m - Ne^2/r = H_0(\theta)$$

Assume now that an electromagnetic field with vector and scalar potentials $A(t, r|Z), \Phi(t, r|Z)$ i incident upon the atom. Here, Z is a random vector that contains the randomness in the electromagnetic field generated by the antenna, the randomness coming from medium fluctuations as well as fluctuations and noise in the antenna source current due to thermal effects. The perturbed Hamiltonian of the electron is given by

$$H(t) = H(t|Z, \theta) = (-1/2m)(\nabla + ieA(t, r|Z))^2 - Ne^2/r - e\Phi(t, r|Z)$$

$$= H_0(\theta) + V_1(t|Z, \theta) + V_2(t|Z, \theta)$$

where

$$V_1(t|Z, \theta) = (-ie/2m)(divA(t, r|Z) + 2(A(t, r|Z, \theta), \nabla)) - e\Phi(t, r|Z)$$

$$V_2(t|Z, \theta) = e^2A^2(t, r|Z)/2m$$

Note that V_1 is $O(e)$ while V_2 is $O(e^2)$. The density operator of the electron $\rho(t) = \rho(t|Z,\theta)$ satisfies

$$i\partial_t \rho(t) = [H(t), \rho(t)]$$

Writing

$$\rho(t) = \rho_0(t) + \rho_1(t) + \rho_2(t) + O(e^3)$$

where $\rho_1(t)$ is $O(e)$ and $\rho_2(t)$ is $O(e^2)$, we get upto $O(e^2)$ on equating terms of $O(e^0), O(e^1), O(e^2)$ respectively

$$i\partial_t \rho_0 = [H_0, \rho_0],$$

$$i\partial_t \rho_1 = [H_0, \rho_1] + [V_1, \rho_0],$$

$$i\partial_t \rho_2 = [H_0, \rho_2] + [V_1, \rho_1] + [V_2, \rho_0]$$

The solutions are for $t > s$

$$\rho_0(t) = exp(-i(t-s)ad(H_0))(\rho(s)),$$

$$\rho_1(t) = -i \int_s s^t exp(-i(t-u)ad(H_0))([V_1(u), \rho_0(u)])du$$

$$\rho_2(t) = -i \int_s^t exp(-i(t-u)ad(H_0))([V_1(u), \rho_1(u)] + [V_2(u), \rho_0(u)])du$$

This approximate solution can be expressed as

$$\rho(t) = \rho(t|Z,\theta) = T_{t,s}(Z,\theta)(\rho(s))$$

where $T_{t,s}(Z,\theta) = T_{t,s}$ is a random linear operator (quantum dynamical group) acting on the space of density operators. Suppose we have a PVM (projection valued measurement) $\{P_a\}$ with the measurements on this system taken at times $t_1 < t_2 < ... < t_N$, noting the outcome after each measurement and then incorporating the state collapse postulate. Then the joint probability of measuring the outcomes $a_1, ..., a_N$ respectively at these times is given by

$$P(a_1, ..., a_N; t_1, ..., t_N|Z,\theta) =$$

$$Tr(T_{t_N,t_{N-1}}(P_{a_{N-1}}...(P_{a_2}T_{t_2,t_1}(P_{a_1}T_{t_1,0}(\rho(0))P_{a_1}).P_{a_2})...P_{a_{N-1}})P_{a_N})$$

Note that the dependence of this joint probability on Z, θ stems from the fact that the evolution operators T_{t_{k+1},t_k} depend upon these parameters where Z is random with known pdf while θ is unknown and non-random. To estimate θ using the EM algorithm, we let $p_Z(Z)$ denote the pdf of Z and then the pdf of $X = (a_1, ..., s_N)$ is

$$p_X(X|\theta) = \int p(X|Z,\theta)p_Z(Z)dZ$$

Directly maximizing this w.r.t θ gives the MLE of θ. This expression being very complicated in general to maximize, we adopt the EM algorithm to construct a sequence of approximations $\theta(n), n = 1, 2, ...$ to the MLE as

$$\theta(n+1) = argmax_\theta Q(\theta, \theta(n))$$

where

$$Q(\theta, \theta(n)) = \int ln(p(X|Z, \theta))p(Z|X, \theta(n))dZ$$

where

$$p(Z|X, \theta(n)) = p(X|Z, \theta(n))p_Z(Z)/p_X(X)$$

So we can equivalently formulate the EM recursion as

$$\theta(n+1) = argmax_\theta \int ln(p(X|Z, \theta)).p(X|Z, \theta(n))p_Z(Z)dZ$$

The EM algorithm in quantum filtering theory. Consider the HPS QSDE

$$dU(t) = (-(iH + P)dt + LdA(t) - L^* dA(t)^*)U(t), P = LL^*/2$$

The non-demolition measurement process is

$$Y_o(t) = U(t)^* Y_i(t)U(t), Y_i(t) = c.A(t) + \bar{c}.A(t)^*$$

If is clear that

$$dY_o(t) = dY_i(t) - j_t(cL^* + \bar{c}L)dt$$

So this measurement model amounts to taking noisy measurements of the system observable $-(cL^* + \bar{c}L)$ evolved at time t via the HPS dynamics. Now the Belavkin estimate of the system observable X on the system at time t given that the bath is in a coherent state $|\phi(u)>$ has the form

$$d\pi_t(X) = \pi_t(\mathcal{L}X)dt + (\pi_t(M_tX + XM_t^*) - \pi_t(M_t + M_t^*)\pi_t(X))(dY_o(t) - \pi_t(M_t + M_t^*)dt)$$

where M_t is a system observable constructed from L, L^* and $u(t), \bar{u}(t)$. Note that

$$dY_o(t) - \pi_t(M_t + M_t^*)$$

is the differential of a classical Wiener process in the coherent state $|\phi(u)>$. The Belavkin filter can be cast as a filter for the system state $\rho_s(t)$ which can be viewed as a classical random process with values in the space of system density matrices by writing

$$\pi_t(X) = Tr(\rho_s(t)X)$$

Thus, by duality, the Belavkin filter for the state filter is given by

$$d\rho_s(t) = \mathcal{L}^*(\rho_s(t))dt + ((\rho_s(t)M_t + M_t^*\rho_s(t)) - Tr(\rho_s(t)(M_t + M_t^*))\rho_s(t))(dY_o(t) \\ - Tr(\rho_s(t)(M_t + M_t^*))dt)$$

Now suppose that that the system Hamiltonian H_0 and the system Lindblad operator L depend upon unknown non-random parameters θ to be estimated. Then, the generator $\mathcal{L}^*$ as well as M_t will depend upon θ and hence we can write $\mathcal{L}^* = \mathcal{L}^*(\theta), M_t = M_t(\theta)$. Our aim is to estimate this parameter θ by taking measurements on the filtered system state $\rho_s(t)$ at different times. The fact is that we can measure only non-demolition processes like $Y_o(t)$ and not directly the true system state $\rho_{s0}(t)$ which satisfies the GKSL equation. By measuring $Y_o(t)$ continuously, we can prepare our measurement box in the Belavkin filtered state upon which we take PVM measurements at different times $t_1 < t_2 < ... < t_N$. Writing our solution to the Belavkin state as $\rho_s(t, Y_o|\theta), t \geq 0$, our goal is to take PVM measurements $\{P_a\}$ at times $t_1, ..., t_N$ and then estimate θ. We write

$$\rho_s(t, Y_o|\theta) = T_{t,s}(Y_o, \theta)(\rho_s(s)), t > s$$

By making measurements using a PVM $\{P_a\}$ of the evolving filtered state at times $t_1 < t_2 < ... < t_N$, we get a joint probability distribution of the measurement outcomes of the form

$$P(a_1, ..., a_N|Y_o, \theta) =$$

$$Tr(T_{t_N,t_{N-1}}(Y_o, \theta)(P_{a_{N-1}}...(P_{a_2}T_{t_2,t_1}(Y_o, \theta)(P_{a_1}T_{t_1,0}(Y_o, \theta)(\rho(0))P_{a_1}).P_{a_2})...P_{a_{N-1}})P_{a_N})$$

and implementing the ML estimator directly would involve maximizing

$$\int P(a_1, ..., a_N|Y_o, \theta)dP_Y(Y_o)$$

with P_Y being the probability distribution of the process Y_o in the coherent state $|\phi(u)>$. This distribution can be determined using the fact that $Y_o(t) - \int_0^t Tr(\rho_s(\tau)(M_\tau + M_\tau^*))du$ is a standard Wiener process in this coherent state. Specifically, an algorithm for calculating the distribution of Y_o in this coherent state would be recursive: Let δ be the time discretization interval and write approximately the discretized Belavkin filter equation as

$$\rho_s(t+\delta) = \rho_s(t)+[\mathcal{L}^*(\rho_s(t))\delta+((\rho_s(t)M_t+M_t^*\rho_s(t))$$
$$-Tr(\rho_s(t)(M_t+M_t^*))\rho_s(t))(Y_o(t+\delta)-Y_o(t)-Tr(\rho_s(t)(M_t+M_t^*))\delta)$$

Assuming that the joint distribution of $Y_o(s) : s \leq t$ is known, then using the fact that $\rho_s(\tau), \tau \leq t$ is determined completely from $Y_o(s) : s \leq t$, we can calculate the conditional distribution of $\rho_s(t + \delta)$ given $Y_o(s) : s \leq t$ from the above equation by using the fact that conditioned on $Y_o(s) : s \leq t$, the random variable $Y_o(t+\delta) - Y_o(t) - Tr(\rho_s(t)(M_t + M_t^*))$ is $N(0, \delta)$. Then, the conditional distribution of $Y_o(t + 2\delta)$ given $Y_o(s) : s \leq t + \delta$ can be determined by using the fact that

$$Y_o(t + 2\delta) - Y_o(t + \delta) - Tr(\rho_s(t + \delta)(M_{t+\delta} + M_{t+\delta}^*)))$$

conditioned on $Y_o(s) : s \leq t + \delta$ is $N(0, \delta)$ and that $\rho_s(t + \delta)$ is completely determined by $Y_o(s) : s \leq t + \delta$.

By regarding the process $Y_o(.)$ as the Latent random process, the EM algorithm can be used to estimate θ from the PVM measurements taken at times $t_1, ..., t_N$.

14.6 The EM algorithm and large deviation theory

$$p(X|\theta) = \int p(X, Z|\theta)dZ = \int p(X|Z, \theta)p(Z|\theta)dZ$$

This is to be maximized w.r.t θ. Let $q(Z)$ be any pdf. Then,

$$log(p(X|\theta)) = \int log(p(X, Z|\theta)/q(Z))q(Z)dZ + \int log(q(Z)/p(Z|X, \theta))q(Z)dZ$$

Start with some initial estimate θ_0 of θ and set $q(Z) = p(Z|X, \theta_0)$. Then if we maximize

$$\int log(p(X, Z|\theta)/q(Z))q(Z)dZ = \int log(p(X, Z|\theta)/p(Z|X, \theta_0))p(Z|X, \theta_0)dZ$$

w.r.t θ or equivalently maximize

$$\int log(p(X, Z|\theta)p(Z|X, \theta_0)dZ$$

w.r.t θ, then the value of the log likelihood function $log(p(X|\theta))$ will only increase. This is because we always have

$$\int q(Z).log(q(Z)/p(Z|X, \theta))dZ \geq 0$$

and when $\theta = \theta_0$ and $q(Z) = p(Z|X, \theta_0)$, this quantity equals zero. Now consider the iteration

$$\theta_1 = argmax_\theta \int log(p(X, Z|\theta))p(Z|X, \theta_0)dZ$$

To evaluate this approximately write

$$\theta_1 = \theta_0 + \delta\theta$$

and expand upto $O(\delta\theta^2)$ to get

$$log(p(X, Z|\theta)) = L(X, Z|\theta) \approx L(X, Z|\theta_0) + L'(X, Z|\theta_0)\delta\theta$$

$$+(1/2)\delta\theta^T L''(X, Z|\theta_0)\delta\theta$$

So by the above maximization process, $\delta\theta$ approximately satisfies

$$(\int L''(X, Z|\theta_0)p(Z|X, \theta_0)dZ)\delta\theta + \int L'(X, Z|\theta_0)p(Z|X, \theta_0)dZ = 0$$

or equivalently,

$$\delta\theta = -[\int L''(X, Z|\theta_0)p(Z|X, \theta_0)dZ]^{-1}.[\int L'(X, Z|\theta_0)p(Z|X, \theta_0)dZ]$$

By applying this to an iid sequence $(X,Z) = ((X_n, Z_n), n = 1,2,...,N)$ and using

$$L(X,Z|\theta) = \sum_{n=1}^{N} L(X_n, Z_n|\theta)$$

we can derive using Cramer's theorem, the large deviation rate function of $\delta\theta$. This procedure can be generalized to include the correction to the parameter estimate after several iterations.

14.7 Kolmogorov-Smirnov statistics

Let $B(t), t \geq 0$ be standard Brownian motion and let $a < b$. Let $x \in [a,b]$ and $[c,d] \subset [a,b]$. For $k \geq 0$, let $A(k,1)$ denote the event that the process starting at x first hits the level a (before hitting b) and then crosses the interval $[a,b]$ exactly k times in the time interval $[0,T]$ and finally at time T $B(T) \in [c,d]$. Likewise, let $A(k,2)$ denote the event that the process starting at x first hits the level b (before hitting a) and then crosses the interval $[a,b]$ exactly k times and finally at time T, $B(T) \in [c,d]$. It is clear that if A denotes the event that the process stays within the interval $[a,b]$ throughout the duration $[0,T]$ and $B(T) \in [c,d]$, then

$$A = \{B(T) \in [c,d]\} - \bigcup_{k\geq 0}(A(k,1) \cup A(k,2))$$

By application of the reflection principle, the probability of $\bigcup_{k\geq 0}(A(k,1)\cup A(k+1,2))$ (which is the event that the process hits a in $[0,T]$ at least once and then at time T lands in $[c,d]$) equals the probability of the event $B(T) \in [2a-d, 2a-c]$ (This follows by applying one reflection at a).

The probability of $\bigcup_{k\geq 1}(A(k,1) \cup A(k+1,2))$ (which is the event that the process hits a at least once and then makes at least one crossing of $[a,b]$ after hitting a and at time T lands in $[c,d]$) equals the probability of the event $\{B(T) \in\in [2(2a-b)) - (2a-c), 2(2a-b) - (2a-d)]\} = \{B(T) \in [2(a-b)+c, 2(a-b)+d]$ (This follows by applying one reflection at a and then one reflection at b).

This is understood more easily by the following argument: Let $T_1 B(t)$ denote the reflection of $B(t)$ after the time $\tau_1 = min(t \geq 0 : B(t) = a\}$. Let $T_2 B(t)$ denote the reflection of $T_1 B(t)$ after the time $\tau_2 = min(t \geq \tau_1 : B(t) = b\} = min(t \geq \tau_1 : T_1 B(t) = 2a - b\}$. Let $T_3 B(t)$ denote the reflection of $T_2 B(t)$ after the time $\tau_3 = min(t \geq \tau_2 : B(t) = a\} = min(t \geq 0 : T_2 B(t) = 2(2a-b) - a\} = min(t \geq 0 : T_2 B(t) = 3a - 2b\}$. In general, we find that for $l = 0,1,...,$ $\tau_{2l+1} = min(t \geq \tau_{2l} : B(t) = a\} = min(t \geq \tau_{2l} : T_{2l} B(t) = a + 2l(a-b)\}$ and $\tau_{2l+2} = min(t \geq \tau_{2l+1} : B(t) = b\} = min(t \geq \tau_{2l+1} : T_{2l+1} B(t) = a + (l+1)(a-b)\}$. Equivalently, $\tau_{l+1} = min(t \geq 0 : T_l B(t) = a + l(a-b)\}$.

Then probability of the event $\bigcup_{k\geq 2}(A(k,1)\cup A(k+1,2))$ (which is the event that the process hits a at least once and then makes at least two crossings of

$[a, b]$ after hitting a and at time T lands in $[c, d]$) equals the probability of the event

$$\{B(T) \in [2(2(2a-b)-a)) - (2(a-b)+d)), 2(2(2a-b)-a)) - (2(a-b)+c))]\}$$

$$= \{B(T) \in [4a-2b-d, 4a-2b-c]\}$$

(This follows by applying one reflection at a, then one reflection at $b \to 2a-b$ and then one reflection at $a \to 2(2b-a)-a$).

Likewise, the probability of the event $\bigcup_{k\geq 3}(A(k,1) \cup A(k+1,2))$ (which is the event that the process hits a at least once, then makes at least three crossings of $[a, b]$ after hitting a and at time T lands in $[c, d]$) equals the probability of the event

$$\{B(T) \in [2(2(2(2a-b)-a)-(2a-b))-(4a-2b-c),$$

$$2(2(2(2a-b)-a)-(2a-b))-(4a-2b-d)]$$

$$= \{B(T) \in [4(a-b)+c, 4(a-b)+d]\}$$

In general, after the lth reflection at time τ_l, let a go to $a(l)$, b to $b(l)$, c to $c(l)$ and d to $d(l)$. Note that the process B goes after the lth reflection to $T_l B(t)$. The first reflection is at τ_1 and hence $a(0) = a(1) = a, b(0) = b, b(1) = 2a-b$. For $l = 0, 1, 2, ...$, the $(2l+1)^{th}$ reflection is at time τ_{2l+1} and $T_{2l+1}B$ is obtained by reflecting $T_{2l}B$ around the level $a(2l)$. For $l = 1, 2, ...$, the $2l^{th}$ reflection is at time τ_{2l} and $T_{2l}B$ is obtained by reflecting $T_{2l-1}B$ around the line $b(2l-1)$. We thus observe the recursion

$$b(2l+1) = 2a(2l)-b(2l), a(2l+1) = a(2l), b(2l) = b(2l-1), a(2l)$$
$$= 2b(2l-1)-a(2l-1), l \geq 1$$

From these equations, we derive

$$b(2l+2) = 2a(2l) - b(2l), a(2l) = 2b(2l) - a(2l-2), l = 1, 2, ...$$

with the initial conditions

$$a(0) = a, b(0) = b, a(2) = 2b-a, b(2) = 2a-b$$

From the above, we again derive

$$(b(2l+2) + b(2l))/2 = 2b(2l) - (b(2l) + b(2l-2))/2$$

which is the same as

$$b(2l+2) = 2b(2l) - b(2l-2)$$

which has the general solution

$$b(2l) = \alpha l + \beta$$

Putting in the initial conditions

$$b(0) = b, b(2) = 2a-b$$

gives

$$\alpha = 2(a-b), \beta = b$$

so that

$$b(2l) = b + 2l(a-b), l = 0, 1, 2, ...$$

Then, from the above equations, we have

$$a(2l) = (b(2l+2) + b(2l))/2 = b + (2l+1)(a-b), l = 0, 1, 2, ...$$

and further,

$$b(2l+1) = 2a(2l) - b(2l) = b + 2(l+1)(a-b)$$

$$a(2l+1) = a(2l) = b + (2l+1)(a-b)$$

Again, we find that

$$c(2l+1) = 2a(2l) - c(2l), c(2l+2) = 2b(2l+1) - c(2l+1)$$

from which we deduce that

$$c(2l+2) = 2b(2l+1) - 2a(2l) + c(2l) = c(2l) + 2(a-b)$$

and thus, using the initial condition $c(0) = 0$, we get

$$c(2l) = c + 2l(a-b), l = 0, 1, 2, ...$$

and hence

$$c(2l+1) = 2a(2l) - c(2l) = 2b - c + 2(l+1)(a-b), l = 0, 1, 2, ...$$

Let $m \geq 0$. The event $W(m,1) = \bigcup_{k \geq m}(A(k,1) \cup A(k+1,2))$ is the event that the process hits the level a in $[0,T]$, then after than makes at least m crossings of the interval $[a,b]$ and finally at time T lands in $[c,d]$. When $W(m,1)$ occurs, the process makes at least m crossings of $[a,b]$ after hitting a, which means that $m+1$ reflections take place in computing the probability of $W(1,m)$.

Thus, by the reflection principle, the probability of this event $W(2m,1)$ is

$$P(B(T) \in [d(2m+1), c(2m+1)])$$

and likewise the probability of $W(2m+1,1)$ is

$$P(B(T) \in [c(2m+2), d(2m+2)])$$

Likewise, let $\sigma_1 = min(t \geq 0 : B(t) = b)$ and for $l \geq 1$, let $\sigma_{2l} = min(t \geq \sigma_{2l-1} : B(t) = a)$ and $\sigma_{2l+1} = min(t \geq 0 : B(t) = \sigma_{2l}\}$. Again, let for any $c \in \mathbb{R}$, $c'(l)$ denote the point c after it undergoes l reflections at times $\sigma_1, ..., \sigma_l$. Then, as before, we have for $l \geq 0$,

$$c'(2l+1) = 2a - c + 2(l+1)(b-a), c'(2l) = c + 2l(b-a)$$

As a simple check we get $c'(0) = c, c'(1) = 2a - c + 2(b - a) = 2b - c$. Thus, letting $W(m, 2) = \bigcup_{k \geq m}(A(k+1, 1) \cup A(k, 2))$, we first observe that that for $m \geq 0$, $W(m, 2)$ is the event that the process hits b in $[0, T]$, then crosses $[a, b]$ at least m times and finally at time T, lands in $[c, d]$. We find as earlier that the probability of $W(2m, 2)$ equals

$$P(B(T) \in [d'(2m+1), c'(2m+1)])$$

while the probability of $W(2m+1, 2)$ is

$$P(B(T) \in [c'(2m+2), d'(2m+2)])$$

Now we observe that the probability that the process stays in $[a, b]$ throughout the entire duration $[0, T]$ and finally at T, lands up in $[c, d]$ is given by

$$P(B(T) \in [a, b]) - P(W(0, 1) \cup W(0, 2))$$

Now, define

$$B(m, 1) = \bigcup_{k \geq m} A(k, 1), B(m, 2) = \bigcup_{k \geq m} A(k, 2), m \geq 0$$

Then, $B(m, 1)$ and $B(r, 2)$ are disjoint for each $m, r \geq 0$ and we can equivalently write

$$P(W(0, 1) \bigcup W(0, 2)) = P(B(0, 1)) + P(B(0, 2))$$

$$P(B(0, 1)) + P(B(0, 2)) = lim_{N \to \infty} \sum_{k=0}^{N} (-1)^m (P(B(m, 1)) + P(B(m, 2))$$

$$P(W(m, 1)) = P(B(m, 1)) + P(B(m+1, 2)), P(W(m, 2))$$
$$= P(B(m+1, 1)) + P(B(m, 2))$$

Then,

$$P(B(0, 1) + P(B(0, 2)) =$$

$$lim_{N \to \infty} \sum_{m=0}^{N} [(-1)^m (P(B(m, 1) +$$

$$P(m+1, 2)) + (-1)^m (P(B(m+1, 1) + P(B(m, 2))]$$

$$= lim_{N \to \infty} \sum_{m=0}^{N} [(-1)^m (P(W(m, 1) + P(W(m, 2))]$$

where we have used the fact that $P(B(m, 1)), P(B(m, 2)) \to 0$ as $m \to \infty$ because a continuous function cannot make infinite number of crossings of a finite interval in finite time. This can be seen from the definition or a continuous process $f(t)$: For any $\epsilon > 0$ there exists a $\delta > 0$ such that $|t - s| < \delta$ implies $|f(t) - f(s)| < \epsilon$ where $t, s \in [0, T]$. divide $[0, T]$ into $N = [T/\delta] + 1$ disjoint intervals all but one of length δ and the remaining one of length $\leq \delta$. Then, if $\epsilon < b - a$, it is clear that $f(t)$ will make only a finite number of crossings of

$[a, b]$ in the time interval $[0, T]$. More precisely, it will make no more than $2N$ crossings. We've also made use of the fact that

$$W(m,1) = B(m,1) \cup B(m+1,2), B(m,1) \cap B(m+1,2) = \phi,$$

$$W(m,2) = B(m+1,1) \cup B(m,2), B(m+1,1) \cap B(m,2) = \phi,$$

14.8 Quantum transmission lines, LDP problems

The line differential equations are

$$-\partial_z v(t,z) = L\partial_t i(t,z) + Ri(t,z),$$

$$-\partial_z i(t,z) = C\partial_t v(t,z) + Gv(t,z)$$

The line has length d. Expanding the line voltage and current as a spatial Fourier series,

$$v(t,z) = \sum_n v_n(t)exp(-2\pi inz/d), i(t,z) = \sum_n i_n(t)exp(-2\pi inz/d)$$

we get

$$2\pi inv_n(t)/d = Li_n'(t) + Ri_n(t), 2\pi ini_n(t)/d = Cv_n'(t) + Gv_n(t)$$

For each n, these equations describe a damped harmonic oscillator as can be seen by eliminating $i_n(t)$ to get

$$(2\pi in/d)^2 v_n(t) = L(Cv_n'' + Gv_n') + R(Cv_n' + Gv_n)$$

or equivalently,

$$LCv_n'' + (LG + RC)v_n' + (RG + (2\pi n/d)^2)v_n = 0$$

We leave it as an exercise to likewise eliminate v_n and derive the corresponding damped harmonic oscillator differential equation for i_n. It follows from basic Heisenberg matrix mechanics that the same damped harmonic differential equations can be derived quantum mechanically from

$$dX(t)/dt = i[H, X(t)] - (1/2)\sum_n (L_n L_n^* X + XL_n L_n^* X - 2L_n X L_n^*)$$

$$= i[H, X] - (1/2)\sum_n (L_n[L_n^*, X] + [X, L_n]L_n^*)$$

where

$$L_n = c(n)a(n) + d(n)a(n)^*, H = \sum_n \omega(n)a(n)^* a(n)$$

with

$$[a(n), a(m)^*] = \delta[n-m], [a(n), a(m)] = 0$$

$c(n), d(n)$ are complex numbers which in this problem of transmission lines are independent of n.

The LDP problem: Observe that the state satisfies the dual Lindblad equation

$$\rho'(t) = -i[H, \rho(t)] - (1/2)\sum_n (L_n L_n^* \rho(t) + \rho(t) L_n L_n^* - 2L_n^* \rho(t) L_n)$$

We use the Glauber-Sudarshan representation to expand

$$\rho(t) = \int \rho(t,u)|\phi(u)><\phi(u)|du d\bar{u}$$

where

$$u = ((u(n))_{n=1}^{\infty}, u(n) \in \mathbb{C}$$

and $|\phi(u)>$ is the coherent state for the system of oscillators:

$$a(n)|\phi(u)>= u(n)|\phi(u)>, a(n)^*|\phi(u)>= \frac{\partial}{\partial u(n)}|\phi(u)>$$

to derive the pde satisfied by the complex valued function $\rho(t,u)$. We then assume the damping to be small which amounts to saying that the Lindblad operator terms L_n are weighed by a small perturbation parameter ϵ. Then solve this pde for $\rho(t,u)$ perturbatively and calculate using it the approximate probability of a transition between to stationary states of the harmonic oscillator Hamiltonian in the absence of the Lindblad terms and evaluate the rate at which this transition probability converges to zero as the Lindblad parameter $\epsilon \to 0$. Observe that

$$H|\phi(u)>= \sum_n \omega(n)u(n)a(n)^*|\phi(u)>= \sum_n \omega(n)u(n)\frac{\partial}{\partial u(n)}|\phi(u)>$$

Thus,

$$[H, \rho(t)] = \int \rho(t,u)[H, |\phi(u)><\phi(u)|]du d\bar{u}$$

$$= \sum_n \omega(n) \int \rho(t,u)(u(n)\frac{\partial}{\partial u(n)} - \bar{u}(n)\frac{\partial}{\partial u(n)})(|\phi(u)><\phi(u)|)du d\bar{u}$$

$$= \sum_n \omega(n) \int [-\frac{\partial}{\partial u(n)}(u(n)\rho(t,u)) + \frac{\partial}{\partial \bar{u}(n)}(\bar{u}(n)\rho(t,u))]|\phi(u)><\phi(u)|du d\bar{u}$$

Chapter 15

More Tools in Probability, Electron Mass in the Presence of Gravity and Electromagnetic Radiation, More on LDP in Quantum Field Theory, Non-Abelian Gauge Field Theory and Gravitation

1.**Weak convergence of probability measures on a metric space**

2.**The Lindberg conditions and the central limit theorem**

3.**The Stone-Weierstrass theorem and its application to the proof of Prohorov's compactness theorem**

Let L be a closed lattice of $C(X,\mathbb{R})$ where X is a compact metric space. Suppose L has the property that given any $a,b \in \mathbb{R}$ and $x \neq y$ in X there exists an $f \in L$ such that $f(x) = a, f(y) = b$. Then $L = C(X,\mathbb{R})$. To prove this, let $f \in C(X,\mathbb{R})$ be arbitrary and $\epsilon > 0$. Choose and fix an $x \in X$. For any $y \neq x$ define

$$G_y - \{z : f_y(z) < f(z) + \epsilon\}$$

where $f_y \in L$ is such that $f_y(x) = f(x), f_y(y) = f(y)$. Then G_y is open in X and further contains y and also x. Thus, $\{G_y : y \in X\}$ is an open cover of X

and by compactness of X, there exists a finite subset $\{y_1, ..., y_N\}$ of X such that $y_j \neq x \forall j$ and

$$X = \bigcup_{j=1}^{N} G_{y_j}$$

Let

$$g_x = min(f_{y_1}, ..., f_{y_N})$$

Then $g_x \in L$ since L is a lattice . Further, $f_{y_j}(x) = f(x) > f(x) - \epsilon$ for all j an hence,

$$g_x(x) > f(x) - \epsilon \forall x \in X$$

Define the open sets

$$H_x = \{z : g_x(z) > f(z) - \epsilon\}$$

Then $x \in H_x$ and hence

$$X = \bigcup_{x \in X} H_x$$

so that by compactness of X, there is a finite subset $\{x_1, ..., x_M\}$ of X such that

$$X = \bigcup_{j=1}^{M} H_{x_j}$$

Let

$$g = max(g_{x_j} : j = 1, 2, ..., M)$$

Then again $g \in L$ since L is a lattice. Further, $x \in H_{x_j}$ implies $g(x) \geq g_{x_j}(x) > f(x) - \epsilon$. Thus

$$g(x) > f(x) - \epsilon \forall x \in X$$

and further suppose $y \in X$ is arbitrary. Then, $y \in G_{y_j}$ for some j. Then, $g_x(y) \leq f_{y_j}(y) < f(y) + \epsilon$ forall $x \in X$. In particular, $g_{x_k}(y) < f(y) + \epsilon, k = 1, 2, ..., M$ and therefore,

$$g(y) < f(y) + \epsilon$$

Thus,

$$g(y) < f(y) + \epsilon \forall y \in X$$

Thus, we have proved that

$$f(x) - \epsilon < g(x) < f(x) + \epsilon \forall x \in X$$

ie

$$\| f - g \| < \epsilon$$

which proves that L is dense in $C(X, \mathbb{R})$ and since L is closed, $L = C(X, \mathbb{R})$.

15.1 On the amount of mass that an electron can get from the background electromagnetic and gravitational fields

The relevant equations are the Maxwell, Dirac and the Einstein field equations:

$$(F^{\mu\nu}\sqrt{-g})_{,\nu} = -J^\mu\sqrt{-g}$$

$$[\gamma^\mu(x)(i\partial_\mu + eA_\mu(x) + i\Gamma_\mu(x)) - m]\psi(x) = 0$$

where

$$\gamma^\mu(x) = \gamma^a e_a^\mu(x) = \gamma^a(\delta_a^\mu + \delta e_a^\mu(x))$$
$$= \gamma^\mu + \delta\gamma^\mu(x)$$

where

$$\delta\gamma^\mu(x) = \gamma^a \delta e_a^\mu(x)$$

Note that

$$\Gamma_\mu(x) = (1/2)e^{a\nu}(x)e_{a\nu:\mu}(x)$$
$$= (1/2)e^{a\nu}(x)(e_{a\nu,\mu}(x) - \Gamma^\rho_{\nu\mu}(x)e_{a\rho}(x))$$

The expression for the electromagnetic current J^μ is derived from the Dirac Lagrangian

$$L_D = \psi^*\gamma^0[\gamma^\mu(x)(i\partial_\mu + eA_\mu + i\Gamma_\mu) - m]\psi(x)\sqrt{-g}$$

Thus,

$$J^\mu(x)\sqrt{-g(x)} = \partial L_D/\partial A_\mu = e\psi(x)^*\gamma^0\gamma^\mu(x)\psi(x)\sqrt{-g(x)}$$

or equivalently

$$J^\mu = e\psi(x)^*\gamma^0\gamma^\mu(x)\psi(x) = e\psi(x)^*\alpha^a\psi(x)e_a^\mu(x)$$

where

$$\alpha^a = \gamma^0\gamma^0$$

and the Dirac α matrices which are Hermitian. The Dirac equation can be expressed as

$$[(\gamma^\mu + \delta\gamma^\mu(x))(i\partial_\mu) - m]\psi(x)$$
$$= -e(\gamma^\mu + \delta\gamma^\mu(x))(A_\mu(x) + i\Gamma_\mu(x))\psi(x)$$

or

$$[i\gamma^\mu\partial_\mu - m]\psi(x) =$$
$$-(i\delta\gamma^\mu(x)\partial_\mu - m)\psi(x)$$
$$-e(\gamma^\mu + \delta\gamma^\mu(x))(A_\mu(x) + i\Gamma_\mu(x))\psi(x)$$

From this we get for the electron propagator

$$S(x,y) = < T(\psi(x)\psi(y)^*) >$$

the equation

$$[i\gamma^\mu\partial_\mu - m]S(x,y) =$$
$$i\gamma^0 < [\psi(t,r),\psi^*(t,r')] > \delta(t-t')+$$
$$-i < T(\delta\gamma^\mu(x)\partial_\mu - m)\psi(x).\psi(y)^*) >$$
$$-e < T(\gamma^\mu + \delta\gamma^\mu(x))(A_\mu(x) + i\Gamma_\mu(x))\psi(x).\psi(y)^*) >$$

We observe that the canonical Dirac momentum field conjugate to the position field $\psi(x)$ is given by

$$\pi(x)^T = \partial L_D/\partial\partial_0\psi(x) =$$
$$i\psi^*(x)\gamma^0\gamma^0(x)\sqrt{-g(x)}$$

or equivalently, noting that

$$e = det(e^a_\mu(x)) = \sqrt{-g(x)}$$

we get

$$\pi(x)^T = i\psi(x)^*\alpha^a e^0_a(x)e(x), \alpha^a = \gamma^0\gamma^a$$

Thus, the canonical equal time anticommutation relations

$$\{\psi(t,r),\pi(t,r')^T\} = i\delta^3(r-r')$$

can be expressed as

$$\{\psi(t,r),\psi(t,r')^*\} = \alpha^a e^0_a(t,r)e(t,r)\delta^3(r-r')$$

15.2 electron propagator corrections in the presence of quantum noise

[1] Dirac Hamiltonian in a radial potential, application of Large deviation theory to computing the statistics of the quantum average of an observable in the presence of Hudson-Parthasarathy noise; corrections to the electron and photon propagator in the presence of Hudson-Parthasarathy noise.

15.3 Large deviation problems for the Schrodinger and Dirac noisy channels

The Dirac matrices are

$$\alpha_r = \begin{pmatrix} 0 & \sigma_r \\ \sigma_r & 0 \end{pmatrix}, r = 1,2,3$$

$$\beta = \begin{pmatrix} I & 0 \\ 0 & -I \end{pmatrix}$$

It is straightforward to verify using

$$\sigma_r \sigma_s + \sigma_s \sigma_r = 2\delta_{rs}$$

that

$$\alpha_r \alpha_s + \alpha_s \alpha_r = 2\delta_{rs}, \alpha_r \beta + \beta \alpha_r = 0$$

Hence, we can using this representation express Dirac's equation in the potential $V(\mathbf{r})$ as

$$((\alpha, P) + \beta m - eV)\psi = E\psi$$

This expands to give with

$$\psi = \begin{pmatrix} \phi \\ \chi \end{pmatrix}$$

$$(\sigma, P)\chi + m\phi - eV\phi = E\phi, (\sigma, P)\phi - m\chi - eV\phi = E\chi$$

or equivalently,

$$(E + eV - m)\phi = (\sigma, P)\chi, (E + eV + m)\chi = (\sigma, P)\phi$$

The corresponding time dependent Dirac equation in a time varying random potential $V(t, r)$ is given by

$$(i\partial_t - eV(t, r) - m)\phi(t, r) = (\sigma, -i\nabla)\chi(t, r)$$

$$(i\partial_t + eV(t, r) + m)\chi(t, r) = (\sigma, -i\nabla)\phi(t, r)$$

Apart from the random scalar potential V, we can also have a random vector potential $\mathbf{A}(t, r)$ in which case, the time varying Dirac equation gets further generalized to

$$(i\partial_t - eV(t, r) - m)\phi(t, r) = (\sigma, -i\nabla)\chi(t, r) + e(\sigma, A(t, r))\chi(t, r)$$

$$(i\partial_t + eV(t, r) + m)\chi(t, r) = (\sigma, -i\nabla)\chi(t, r) + e(\sigma, A(t, r))\phi$$

If this noisy electromagnetic field is quantum noise so that

$$A(t, r) = L(a, b, r)d\Lambda_b^a(t)/dt, , divA = -\partial_t V$$

so that

$$V = (-divL(a, b, r))\Lambda_b^a(t)$$

the Dirac equation becomes

$$dU(t) = (-i((\alpha, -i\nabla) + \beta.m)dt$$

$$+e^2 Q(a, b, r)d\Lambda_b^a(t) - ieL(a, b, r)d\Lambda_b^a(t) + ieG(a, b, r)\Lambda_b^a(t)dt)U(t)$$

where

$$G(a, b, r) = divL(a, b, r)$$

$Q(a,b,r)$ is the quantum Ito correction term and is obtained using the quantum Ito formula

$$(L(a,b,r)d\Lambda_b^a(t))^*(L(c,d,r)d\Lambda_d^c(t)) = \bar{L}(a,b,r).L(c,d,r)d\Lambda_a^b(t)d\Lambda_d^c(t)$$

$$= \bar{L}(a,b,r)L(c,d,r)\epsilon_d^b d\Lambda_a^c(t) = Q(c,a,r)d\Lambda_a^c(t)$$

so that

$$Q(c,a,r) = \epsilon_d^b \bar{L}(a,b,r)L(c,d,r)$$

with summation over the repeated indices b,d being implied. Note that since $A(t,r)$ must be a Hermitian operator, we must have

$$[L(a,b,r)d\Lambda_b^a(t)]^* = L(b,a,r)d\Lambda_s^b(t)$$

ir equivalently,

$$\bar{L}(a,b,r) = L(b,a,r)$$

We can express this noisy Dirac equation as

$$dU(t) = (-i((\alpha,-i\nabla)+\beta.m)dt+(e^2Q(a,b,r)$$
$$-ieL(a,b,r))d\Lambda_b^a(t)+ieG(a,b,r)\Lambda_b^a(t)dt)U(t)$$

In the presence of a quantum electromagnetic field $A_q(t,r), V_q(t,r)$ and a quantum gravitational field in addition to this quantum noisy electromagnetic field, we would have to reformulate this equation as

$$\gamma^\mu(x)(i\partial_\mu + eA_\mu(x) + i\Gamma_\mu(x))\psi(x) = 0$$

where A_μ is the sum of the quantum electromagnetic four potential and the quantum noisy electromagnetic four potential. This could be expressed as

$$i\gamma^0(x)\partial_t U(t) = (-i\gamma^r(x)\partial_r - e\gamma^\mu(x)A_\mu(x) - i\gamma^\mu(x)\Gamma_\mu(x))U(t)$$

where

$$\gamma^0(x) = \gamma^a e_a^0(x)$$

so that

$$(\gamma^0(x))^2 = \gamma^a\gamma^b e_a^0 e_b^0 = \eta^{ab}e_a^0 e_b^0 = g^{00}(x)$$

so that the above equation can be expressed as

$$\partial_t U(t) = (g^{00}(x))^{-1}(-\alpha^r(x)\partial_r + ie\alpha^\mu(x)A_\mu(x) - \alpha^\mu(x)\Gamma_\mu(x))U(t)$$

where

$$\alpha^\mu(x) = \gamma^0(x)\gamma^\mu(x)$$

In this formula, the quantum Ito correction terms have not yet been added. To be specific, have

$$A_{Nm}(x) = L_m(a,b,r)d\Lambda_b^a(t)/dt,\ A_{N0}(x) = -L_{m,m}(a,b,r)\Lambda_b^a(t),$$

$$A_\mu(x) = A_{q\mu}(x) + A_{N\mu}(x)$$

where A_q denotes the quantum electromagnetic field and A_N denotes the quantum noisy electromagnetic field. We have to ensure that the above equation defines a unitary evolution which means that in the absence of noise, the operator

$$H(t) = (g^{00}(x))^{-1}(-\alpha^r(x)i\partial_r - e\alpha^\mu(x)A_{q\mu}(x) - i\alpha^\mu(x)\Gamma_\mu(x))$$

must be a Hermitian operator. If it is not, then we must replace it by one half times its sum with its Hermitian adjoint.

15.4 Central limit theorem for martingales

Let $X(n), n \geq 0$ be a Martingale adapted to the filtration $\mathcal{F}_n$. Define $Y(n) = X(n)-X(n-1)$ and assume that $|Y(n)| \leq K < \infty \forall n$. Let $\sigma(n)^2 = \mathbb{E}(Y(n)^2|\mathcal{F}_{n-1})$. Note that $\mathbb{E}(Y(n)|\mathcal{F}_{n-1}) = 0$. Assume that $\sum_n \sigma(n)^2 = \infty$. For any positive integer k, let $\tau(k) = min(N \geq 1 : \sum_{n=1}^N \sigma(n)^2 > k)$. Note that since $\sum_n \sigma(n)^2 = \infty$, it follows that $\tau(k)$ is a finite r.v, ie $\tau(k) < \infty$ (a.e.). Define

$$Y(n,k) = \chi_{\tau(n)\geq k}Y(k)/\sqrt{n}, \mathcal{F}_{nk} = \mathcal{F}_k$$

Consider

$$\sigma(n,k)^2 = \mathbb{E}(Y(n,k)^2|\mathcal{F}_{n,k-1})$$

Note that the event $\{\tau(n) \geq k\}$ occurs iff $\{\sum_{r=1}^m \sigma(r)^2 < n\}$ occurs for each $m = 1, ..., k-1$. Since $\sigma(r)^2$ is $\mathcal{F}_{r-1}$-measurable, it follows then that $\{\tau(n) > k\}$ is $\mathcal{F}_{k-2}$ measurable and hence

$$\mathbb{E}(Y(n,k)|\mathcal{F}_{n,k-1}) = 0$$

and

$$\sigma(n,k)^2 = \chi_{\tau(n)\geq k}\mathbb{E}(Y(k)^2|\mathcal{F}_{k-1})/n = \chi_{\tau(n)\geq k}\sigma(k)^2/n$$

We now observe that

$$\sum_{k=1}^{\infty}\sigma(n,k)^2 - \sum_{k=1}^{\tau(n)}\sigma(k)^2/n \geq 1$$

and also

$$\sum_{k=1}^{\infty}\sigma(n,k)^2 = \sum_{k=1}^{\tau(n)}\sigma(k)^2/n$$

$$= \sum_{k=1}^{\tau(n)-1}\sigma(k)^2/n + \sigma(\tau(n))^2/n \leq 1 + K^2/n$$

Thus,

$$lim_{n\to\infty}\sum_{k=1}^{\infty}\sigma(n,k)^2 = 1$$

Further we find that since $|Y(n,k)| \leq K/\sqrt{n}$, it follows that

$$\sum_{k=1}^{\infty} \mathbb{E}(Y(n,k)^2 \chi_{|Y(n,k)|>\epsilon})$$

$$\leq \mathbb{E}\sum_{k=1}^{\infty} \mathbb{E}(Y(n,k)^2|\mathcal{F}_{n,k-1})\chi_{K/\sqrt{n}>\epsilon}$$

$$= \mathbb{E}\sum_{k=1}^{\tau(n)} (\sigma(k)^2/n)\chi_{K/\sqrt{n}>\epsilon}$$

$$\leq (1+K^2/n)\chi_{K/\sqrt{n}>\epsilon} \to 0, n \to \infty$$

In other words, we have proved the generalized Lindeberg conditions:

$$lim_n \sum_{k\geq 1} \sigma(n,k)^2 = 1,$$

$$lim_n \sum_{k=1}^{\infty} \mathbb{E}(Y(n,k)^2 \chi_{|Y(n,k)|>\epsilon}) = 0$$

for any $\epsilon > 0$ where

$$\sigma(n,k)^2 = \mathbb{E}(Y(n,k)^2|\mathcal{F}_{n,k-1})$$

Without loss of generality, we can assume that

$$\sum_k \sigma(n,k)^2 = 1 \forall n$$

since in proving the CLT, we are anyway going to take the limit. Fix n and consider

$$S(n,m) = \sum_{k=1}^{m} Y(n,k), S(n) = lim_{m\to\infty} S(n,m)$$

We have

$$S(n,m) = S(n,m-1) + Y(n,m)$$

$$|\mathbb{E}(exp(itS(n)) - exp(-t^2/2))| \leq exp(t^2/2)|\mathbb{E}(exp(itS(n))exp(t^2/2) - 1)|$$

and

$$exp(itS(n))exp(t^2/2) - 1$$

$$= \sum_{m=0}^{\infty} (exp(itS(n,m) + t^2\Sigma_m/2) - exp(itS(n,m-1) + t^2\Sigma_{m-1}/2))$$

where

$$S(n,-1) = 0, \Sigma_{-1} = 0, \Sigma_m = \sum_{k=1}^{m} \sigma(n,k)^2$$

Thus,

$$|\mathbb{E}exp(itS(n,m)+t^2\Sigma_m/2)-exp(itS(n,m-1)+t^2\Sigma_{m-1}/2)|$$

$$\leq exp(t^2/2)\mathbb{E}|\mathbb{E}[exp(itY(n,m))-exp(-t^2\sigma(n,m)^2/2)|\mathcal{F}_{n,m-1}]|$$

since

$$\Sigma_m\leq\sum_k\sigma(n,k)^2=1,\Sigma_m-\Sigma_{m-1}=\sigma(n,m)^2$$

Now,

$$exp(itY(n,m))=1+itY(n,m)-t^2Y(n,m)^2/2+\theta$$

where

$$|\theta|\leq f(t)(|Y(n,m)|^2,|Y(n,m)|^3)<f(t)(|Y(n,m)|^2\chi_{|Y(n,m)|>\epsilon}+\epsilon.|Y(n,m)|^2)$$

Thus,

$$\sum_m\mathbb{E}(\theta)\leq f(t)\sum_m\mathbb{E}(|Y(n,m)|^2\chi_{|Y(n,m)|>\epsilon}|)$$

$$+f(t)\epsilon.\mathbb{E}\sum_m\sigma(n,m)^2$$

which cannot exceed $f(t)\epsilon$ as $n\to\infty$ because $\sum_m\sigma(n,m)^2$ $=1$ and $\sum_m\mathbb{E}(|Y(n,m)|^2\chi_{|Y(n,m)|>\epsilon}|)\to$ 0. This proves that $\sum_m\mathbb{E}(|\theta|)\to 0, n\to\infty$ since $\epsilon>0$ is arbitrary. Now,

$$\mathbb{E}[exp(itY(n,m))|\mathcal{F}_{n,m-1}]=1+\sigma(n,m)^2t^2/2+\mathbb{E}(\theta|\mathcal{F}_{n,m-1})$$

Likewise,

$$exp(-t^2\sigma(n,m)^2/2)=1-t^2\sigma(n,m)^2/2+\theta'$$

where

$$|\theta'|\leq g(t)\sigma(n,m)^4$$

$$\sum_m|\theta'|\leq g(t)max_k\sigma(n,k)^2$$

since $\sum_m\sigma(n,m)^2=1$. Now,

$$\sigma(n,k)^2=\mathbb{E}(Y(n,k)^2\chi_{|Y(n,k)|<\epsilon})+\mathbb{E}[Y(n,k)^2\chi_{|Y(n,k)|>\epsilon}]$$

$$<\epsilon^2+\mathbb{E}[Y(n,k)^2\chi_{|Y(n,k)|>\epsilon}]$$

so that

$$max_k\sigma(n,k)^2\leq\epsilon^2+\sum_k\mathbb{E}[(Y(n,k)^2\chi_{|Y(n,k)|>\epsilon}]$$

$$\to\epsilon^2,n\to\infty$$

Thus,

$$max_k\sigma(n,k)^2\to 0,n\to\infty$$

15.5 More problems in LDP applied to quantum field theory

[1] Let $\phi(x)$ be a field on $\mathbb{R}^n$ with classical field equations given by a Lagrangian $L(\phi, \phi_{,k})$. In the presence of an external current field $J(x)$, the path integral is

$$Z(J) = \int exp(iS(\phi) + i\int J\phi dx)D\phi$$

Let

$$W(J) = log(Z(J)) - log\int exp(iS(\phi) + i\int J\phi dx)D\phi$$

where

$$S(\phi) = \int Ldx$$

The average of the quantum field ϕ in the presence of the current field J is given by

$$<\phi>_J(x) = Z(J)^{-1}\int \phi(x)exp(iS(\phi) + i\int J\phi dx)D\phi = -i\delta W(J)/\delta J(x)$$

If $J(x)$ is a random classical current field with given statistics, then what is the statistics of the classical field $<\phi>_J(x)$? Now observe that for a classical field $\chi(x)$, if we define the classical current field J_χ by

$$<\phi>_{J_\chi} = \chi$$

and put

$$\Gamma(\chi) = \int J_\chi(x)\chi(x)dx + iW(J_\chi)$$

then

$$\delta\Gamma(\chi)/\delta\chi(x) = J_\chi(x) + \int(\delta J_\chi(y)/\delta\chi(x)))\chi(y)dy + i\int(\delta W(J_\chi)/\delta J(y))(\delta J_\chi(y)/\delta\chi(x))dy$$

$$= J_\chi(x)$$

This is called the quantum equation of motion for the field $\chi(x)$. It is the quantum generalization of the classical equation

$$\delta(S(\phi) + \int J(y)\phi(y)dy)/\delta\phi(x) = J(x)$$

We have an analogue of such a situation in classical probability. Let X be a random variable and $\delta\lambda$ another independent random variable. The moment generating function of X evaluated at $\lambda+\delta\lambda$ taking into account the randomness of $\delta\lambda$ is given by

$$\tilde{M}_X(\lambda) = \mathbb{E}exp((\lambda + \delta\lambda)X)$$

$$= \mathbb{E}(M_X(\lambda + \delta\lambda)) = \int M_X(\lambda + u) f(u) du$$

where f is the probability density of $\delta\lambda$. If $\delta\lambda$ is regarded as weak amplitude noise, then we can write approximately

$$\tilde{M}_X(\lambda) = \sum_{n=0}^{N} M_X^{(r)}(\lambda)\mu_r/r!$$

where

$$\mu_r = \int u^r f(u) du = \mathbb{E}((\delta\lambda)^r)$$

This formula has another kind of significance,namely, while calculating the rate function of a sequence of r.v's of the form $S_n/n, S_n = \sum_{k=1}^n X_k$ where the $X_k' s$ are iid, suppose we make an error by using $\lambda + \delta\lambda$ in place of λ in the preliminary step involving the computation of the logarithmic moment generating function. Then the erroneous rate function for this family will be given by

$$\tilde{I}(x) = sup_\lambda(\lambda x - log(\tilde{M}_X(\lambda)))$$

and evaluating this upto the second moment of moment generating function parameter error $\delta\lambda$, we get

$$\tilde{M}_X(\lambda) = M_X(\lambda) + \mu_1.M_X'(\lambda) + \mu_2 M_X''(\lambda)/2$$

$$log(\tilde{M}_X(\lambda)) = log(M_X(\lambda)) + \mu_1 M_X'(\lambda)/M_X(\lambda) + \mu_2 M_X''(\lambda)/2M_X(\lambda) \\ -(\mu_1^2/2)(M_X'(\lambda)/M_X(\lambda))^2$$

It follows then that the error in the rate function will be approximately $\delta I(x)$ where

$$I(x) = sup_\lambda(\lambda.x - log(M_X(\lambda))$$

$$I(x) + \delta I(x) = \tilde{I}(x) = sup_\lambda(\lambda x - log(M_X(\lambda)) - \mu_1 M_X'(\lambda)/M_X(\lambda) \\ -\mu_2 M_X''(\lambda)/2M_X(\lambda) + (\mu_1^2/2)(M_X'(\lambda)/M_X(\lambda))^2)$$

We leave it as an exercise to evaluate this upto $O(\mu_1, \mu_1^2, \mu_2)$ by showing the additional terms to $I(x)$ that are present.

15.6 Schrodinger and Klein-Gordon equations in quantum field theory based on an infinite dimensional Laplacian operator

Consider the KG Lagrangian for a scalar field ϕ with a nonlinear potential term:

$$L(\phi, \phi_{,\mu}) = (1/2)\partial_\mu\phi.\partial^\mu\phi - m^2\phi^2/2 - V(\phi)$$

The canonical position field is $\phi(x)$ and the canonical momentum field is

$$\pi(x) = \partial L/\partial\partial_0\phi = \partial_0\phi$$

Thus, the Hamiltonian density obtained by applying the Legendre transformation to this Lagrangian is

$$\mathcal{H}(\phi, \nabla\phi, \pi) = \pi.\phi - L = (1/2)(\pi^2 + |\nabla\phi|^2 + m^2\phi^2) + V(\phi)$$

The Hamiltonian is the integral of the Hamiltonian density over the spatial volume:

$$H(\phi, \pi) = \int H(\phi(x), \nabla\phi(x), \pi(x))d^3x$$

$$= \int [(1/2)(\pi(x)^2 + |\nabla\phi(x)|^2 + m^2\phi(x)^2) + V(\phi(x))]d^3x$$

and Schrodinger's equation in infinite dimensions is obtained by assuming that the wave functional $\Psi(t, \phi(r) : r \in \mathbb{R}^3)$ satisfied the differential equation

$$i\frac{\partial\Psi(t,\phi(.))}{\partial t} = H\Psi(t, \phi(.))$$

$$= (1/2)(\int [-\delta^2/\delta\phi(r)^2 + |\nabla\phi(r)|^2 + m^2\phi(r)^2 + 2V(\phi(r))]d^3r)\Psi(t, \phi(.))$$

$$= (1/2)\int [-\delta^2\Psi(t,\phi(.))/\delta\phi(r)^2) + |\nabla\phi(r)|^2\Psi(t,\phi(.)) + m^2\phi(r)^2\Psi(t,\phi(.)) + 2V(\phi(r))\Psi(t,\phi(.))]d^3r$$

We can also hope to solve this infinite dimensional Schrodinger equation in which the Hamiltonian appears as an anharmonically perturbed infinite dimensional quantum Harmonic oscillator using the Feymman path integral method. To do this, we choose a cubic box B of side-length L and expand the position field ϕ as a spatial Fourier series

$$\phi(t,r) = \sum_k c(t,k)exp(ik.r)/L^{3/2}, k = (2\pi/L)(k_x, k_y, k_z), k_x, k_y, k_z \in \mathbb{Z}^3$$

Since ϕ is real, we must impose the restrictions

$$c(t, -k) = \bar{c}(t, k)$$

and then we get

$$\int_B Ld^3x = T_1 + T_2 + T_3 + T_4$$

where

$$T_1 = (1/2)\int_B (\partial_0\phi(t,r))^2d^3r = (1/2)\sum_k |\partial_t c(t,k)|^2,$$

$$T_2 = (-1/2)\int_B |\nabla\phi(t,r)|^2d^3r = (1/2)\sum_k k^2|c(t,k)|^2$$

$$T_3 = (-m^2/2)\int_B \phi(t,r)^2d^3r = (m^2/2)\sum_k |c(t,k)|^2$$

$$T_4 = -\int_B V(\phi(t,r))d^3r = -\int_B V(\sum_k c(t,k)exp(ik.r)))d^3r$$

$$= -\sum_{n,k_1,...,k_n} V(k_1,...,k_n)c(t,k_1)c(t,k_2)...c(t,k_n)$$

We denote this Lagrangian by

$$L(c(t,k), \partial_t c(t,k) : k \in (2\pi/L)\mathbb{Z}^3$$

and now formulate the corresponding Hamiltonian:

$$\pi(t,k) = \partial L/\partial\partial_t c(t,k)) = (1/2)c(t,-k) = (1/2)\partial_t \bar{c}(t,k)$$

$$\pi(t,-k) = \partial L/\partial\partial_t c(t,-k) = (1/2)\partial_t c(t,k) = \bar{\pi}(t,k)$$

Thus the Hamiltonian in spatial Fourier space is given by

$$H(c(t,k), \pi(t,k) : k \in 2\pi\mathbb{Z}^3/L) =$$

$$\sum_k (\pi(t,k)\partial_t c(t,k) + \bar{\pi}(t,k)\partial_t \bar{c}(t,k)) - L =$$

$$= (1/2)\sum_k (|\pi(t,k)|^2 + (k^2+m^2)|c(t,k)|^2) +$$

$$+ \sum_{n,k_1,...,k_n} V(k_1,...,k_n)c(t,k_1)c(t,k_2)...c(t,k_n)$$

Schrodinger's equation in terms of this Hamiltonian now appears as

$$i\frac{\partial\Psi(t,c(.))}{\partial t} =$$

$$H(c(k), -i\frac{\partial}{\partial c(k)} : k \in 2\pi\mathbb{Z}^3/L)\Psi(t,c(.))$$

This Schrodinger equation can be solved via a Feynman path integral using an infinite dimensional Brownian motion with a countable number of components with complex time. The unitary evolution kernel is

$$K(T,c|0,d) = C\int_{(0,d)}^{(T,c)} exp(i\int_0^T (1/2)\sum_k |\partial_t c(t,k)|^2 - i\int_0^T W(c(t,.)dt)Dc$$

where

$$W(c(.)) = (1/2)\sum_k ((k^2+m^2)|c(t,k)|^2) + \sum_{n,k_1,...,k_n} V(k_1,...,k_n)c(t,k_1)c(t,k_2)...c(t,k_n)$$

The stationary states of the Harmonic oscillator component are given by

$$|n >= |n_1, n_2, ... >= |n(k) : k \in firsttwoquadrants >= |\{n(k)\} >$$

where

$$c(k)^*c(k)|n>=n(k)|c(k)>$$

The oscillator coherent states are

$$\phi(u)>=\sum_n(\Pi_k u(k)^{n(k)}/\Pi_k\sqrt{n(k)!})|n>$$

Note that

$$|n>=exp(-\parallel u\parallel^2/)\Pi_k\frac{c(k)^{*n(k)}}{\sqrt{n(k)!}}|0>$$

Then, for k in the first two quadrants

$$c(k)|\phi(u)>-u(k)|phi(u)>,c(k)^*exp(\parallel u\parallel^2/2)|\phi(u)>$$

$$=\frac{\partial}{\partial u(k)}exp(\parallel u\parallel^2/)\phi(u)>$$

$$=exp(\parallel u\parallel^2/2)(\partial/\partial u(k)+\bar{u}(k)/2))|\phi(u)>$$

or equivalently,

$$c(k)^*|\phi(u)>=(\partial/\partial u(k)+\bar{u}(k)/2)|\phi(u)>$$

15.7 Proof of the Prohorov tightness theorem

Let X be a separable metric space. By the Urysohn theorem, X can be embedded inside a compact metric space Y, such that X is dense in Y. Formally, this means that there exists a compact metric space Y and a map $\phi : X \to Y$ such that ϕ is injective, continuous and $\phi^{-1} : \phi(X) \to X$ is also continuous and finally that $\phi(X)$ is dense in Y. Now, since Y is compact, $C(Y)$ is a separable metric space in view of the Stone-Weierstrass theorem. Specifically, Y being compact, ia also separable and hence there exists a countable dense subset $\{x_n : n = 1, 2, ...\}\}$ of Y. Then the set of functions $f_n(x) = d(x_n, x), n \geq 1$ separates the points of Y and hence the algebra $\mathcal{A}$ generated by these functions is dense in $C(Y)$ by the Stone-Weierstrass theorem. Now the set of all functions of the form $1, \sum_{n_1,...,n_K=1}^N c(n_1, ..., n_K)f_{n_1}...f_{n_K}$ with $N, K = 1, 2, ...$ and $c(n_1, ..., n_K)$ rational is countable and also dense in $\mathcal{A}$ anbd this proves that $C(Y)$ is a separable metric space. Without loss of generality, we can assume that X is a dense subset of Y by renaming $\phi(X)$ as X.

Now choose a countable dense subset $\{g_n : n = 1, 2, ...\}$ in $C(Y)$ an let f_n be the restriction of g_n to X. Let M be tight family of probability measures on X, ie, for every $\epsilon > 0$ there exists a compact set $K_\epsilon \subset X$ such that $\mu(K_\epsilon) > 1 - \epsilon) forall \mu \in M$. Then we wish to show that M is compact, ie, given any infinite sequence $\mu_n, n = 1, 2, ...$ in M, there exists a subsequence μ_{n_m} that converges weakly to some probability measure μ on X. For any measure μ on

X, we define the measure $\hat{\mu}$ on Y by $\hat{\mu}(B) = \mu(X \cap B), B \in \mathcal{B}(X)$. It is clear that if μ is a probability measure on X, the $\hat{\mu}$ is also a probability measure on Y. Now for any probabiliy measure ν on Y, define

$$T(\nu) = \{\int g_n d\nu : n \geq 1\}$$

It is clear that $T(\nu)$ is a sequence in $[0, 1]$ provided that we replace g_n by $g_n / \parallel g_n \parallel$. Further if ν and μ are two distinct probability measures on Y, then $T(\nu) \neq T(\mu)$. For suppose $T(\nu) = T(\mu)$. Then

$$\int g_n d\nu = \int g_n d\mu \forall n$$

Let g be any continous function on Y and let $\epsilon > 0$. Then there exists an n and a constant $c(n)$ such that

$$\parallel c(n)g_n - g \parallel < \epsilon$$

It follows then by the triangle inequality for complex numbers that

$$|\int g d\nu - \int g d\mu| \leq \int |g - c(n)g_n| d\nu + \int |g - c(n)g_n| d\mu + |c(n)|| \int g_n d\nu - \int g_n d\mu| < 2\epsilon$$

and since $\epsilon > 0$ is arbitrary, we get

$$\int g d\nu = \int g d\mu$$

Since g is an arbitrary continuous function on Y, it follows that $\nu = \mu$. In fact, let C be any closed subset of Y. Then we define $G_n = \{y : d(y, C) < 1/n\}$. G_n is a sequence of open sets that decreases to C and we can choose a continuous function h_n on Y such that $h_n|_C = 1, h_n|_{G_n^c} = 0$ and $0 \leq h_n \leq 1$. Then

$$\nu(C) \leq \int_C h_n d\nu - \int_C h_n d\mu \leq \mu(G_n)$$

and letting $n \to \infty$ gives us

$$\nu(C) \leq \mu(C)$$

Interchanging the ν and μ in this argument then gives us the result that ν and μ coincide on all the closed subsets of Y. Now define $\mathcal{C}$ to be the collection of all Borel subsets B of Y for which $\nu(B) = \mu(B)$. Then, $\mathcal{C}$ is a σ-field containing all the closed subsets of Y. Thus, $\mathcal{C}$ must coincide with $\mathcal{B}(X)$ and the proof of the claim, namely that T is an injective map is complete.

15.8 Large deviation problems in field measurement analysis

Let $F(x), x \in \mathbb{R}^n$ be a random field. Its pdf depends on a vector parameter θ, we write it as

$$p(F(x) : x \in \mathbb{R}^n|\theta)$$

The aim is to estimate θ from measurements of the field F at a given finite set of points $x_1, ..., x_N \in \mathbb{R}^n$. Owing to instrumental errors, we actually measure the field at the points $x_k + \delta x_k, k = 1, 2, ..., N$ where $\delta x_k, k = 1, 2, ..., N$ are random position errors. Assume that these errors have a joint pdf $f(\delta x_k, k = 1, 2, ..., N)$. Then the joint density of these measurements $F_k = F(x_k + \delta x_k), k = 1, 2, ..., N$ are given by

$$q(F_k, k = 1, 2, ..., N|\theta) = \int p(F_k, x_k+\delta x_k, k = 1, 2, ..., N|\theta) f(\delta x_k, \\ k = 1, 2, ..., N)\Pi_{k=1}^{N} d\delta x_k$$

where $p(F_k, x_k, k = 1, 2, ..., N)$ is the joint density of $F(x_k), k = 1, 2, ..., N$. θ is estinmated from q using the maximum likelihood method. Equivalently, we can estimate it using the EM algorthm as follows. Let θ_0 be a preliminary estimate of θ. Then its improvement θ_1 at the next iteration is given by

$$\theta_1 = argmax_\theta Q(\theta, \theta_0)$$

where

$$Q(\theta, \theta_0) = \int log(p(F_k, x_k+\delta x_k, k = 1, 2, ..., N|\theta))p(\delta x_k, k = 1, 2, ..., N|F_k, \\ k = 1, 2, ..., N, \theta_0 \Pi_k d\delta x_k)$$

where

$$p(\delta x_k, k = 1, 2, ..., N|F_k, k = 1, 2, ..., N, \theta_0)$$

in the above expression can be replaced by

$$p(F_k, k = 1, 2, ..., N|\delta x_k, k = 1, 2, ..., N, \theta_0) f(\delta x_k, k = 1, 2, ..., N) \\ = p(F_k, x_k+\delta x_k, k = 1, 2, ..., N|\theta_0) f(\delta x_k, k = 1, 2, ..., N)$$

Specifically to see how the EM algorithm simplifies the computation, we assume that $F(x), x \in \mathbb{R}^n$ is a Gaussian random field with mean

$$M(x|\theta) = \mathbb{E}(F(x))$$

and covariance

$$C(x, y|\theta) = \mathbb{E}(F(x)F(y)) - M(x|\theta)M(y|\theta)$$

Denote the inverse kernel of C by Q. Specifically, Q satisfies

$$\int Q(x, y|\theta)C(y, z|\theta)dy = \delta(x - z)$$

Then, we have

$$p(F_k, x_k, k = 1, 2, ..., N|\theta) = det(C(x_1, ..., x_N|\theta))^{-1/2}$$

$$exp(-(1/2)) \sum_{k,m=1}^{N} Q_{km}(x_1, ..., x_N|\theta)(F_k - M(x_k|\theta)(F_m - M(x_m|\theta))$$

where

$$C(x_1, ..., x_N|\theta) = ((C(x_i, y_j|\theta)))_{1 \leq i,j \leq N}$$

Thus,

$$log(p(F_k, x_k + \delta x_k, k = 1, 2, ..., N|\theta)) = (-1/2).log(detC(x_1 + \delta x_1, ..., x_N + \delta x_N|\theta))$$

$$-(1/2) \sum_{k,m=1}^{N} Q_{km}(x_1, ..., x_N|\theta)(F_k - M(x_k + \delta x_k|\theta).(F_m - M(x_m + \delta x_m|\theta))$$

15.9 ADM action for quantum gravity and its noisy perturbation with LDP analysis of the solution metric

Embed a three dimensional manifold Σ_t at time t into $\mathbb{R}^4$. We denote the spatial coordinates of Σ_t by (x^a). The space-time coordinates of $\mathbb{R}^4$ into which Σ_t is embedded are denoted by $X^\mu(t, x)$. We define

$$T^\mu = X^\mu_{,0} = \partial X^\mu / \partial t$$

Let n^μ denote the unit normal to Σ_t. Then

$$T^\mu = Nn^\mu + N^\mu$$

is a decomposition of the vector X^μ into a normal component and a tangential component to Σ_t. Here, N is a normalization scalar field. N^μ being tangential to Σ_t, can be expressed as

$$N^\mu = N^a X^\mu_{,a}$$

the sum being over $a = 1, 2, 3$. Note that $\partial/\partial x^a$ are the tangent vectors to Σ_t. By orthogonality of n^μ and N^μ, we must have

$$g_{\mu\nu} n^\mu N^\nu = 0$$

where $g_{\mu\nu}$ is the metric on $\mathbb{R}^4$. Denote by $\tilde{g}_{\mu\nu}$, the metric relative to the coordinates $(x^\mu) = (t, x^a)$ so that $x^0 = t$ defines the surface Σ_t. Then, we have

$$g_{\mu\nu} X^\mu_{,a} X^\nu_{,b} = q_{ab} = \tilde{g}_{ab}, g_{\mu\nu} T^\mu X^\nu_{,b} = \tilde{g}_{0b}, g_{\mu\nu} T^\mu T^\nu = \tilde{g}_{00}$$

The above orthogonality relation can be expressed as

$$g_{\mu\nu}(T^\mu - N^\mu)N^\nu = 0$$

or equivalently as

$$g_{\mu\nu}(T^\mu - Nn^\mu)n^\nu = 0$$

Note that more generally, since n^μ is normal to Σ_t, we also have

$$g_{\mu\nu}(T^\mu - N^\mu)X^\nu_{,b} = 0$$

which can be equivalently expressed as

$$\tilde{g}_{0b} - g_{\mu\nu}X^\mu_{,a}X^\nu_{,b}N^a = 0$$

or equivalently,

$$\tilde{g}_{0b} = q_{ab}N^a = N_b$$

where the spatial metric q_{ab} in Σ_t is used to lower the spatial indices. We wish to show that the metric $g^{\mu\nu}$ in $\mathbb{R}^4$ admits the orthogonal decomposition

$$g^{\mu\nu} = q^{\mu\nu} + n^\mu n^\nu$$

where

$$q^{\mu\nu}n_\nu = 0$$

To this end, we observe that

$$g^{\mu\nu} = \tilde{g}^{\alpha\beta}X^\mu_{,\alpha}X^\nu_{,\beta} = \tilde{g}^{a0}(X^\mu_{,a}T^\nu + X^\nu_{,a}T^\mu) + \tilde{g}^{ab}X^\mu_{,a}X^\nu_{,b} + \tilde{g}^{00}T^\mu T^\nu$$

$$= 2\tilde{g}^{a0}X^\mu_{,a}(N^\nu + Nn^\nu) + \tilde{g}^{ab}X^\mu_{,a}X^\nu_{,b} + \tilde{g}^{00}(N^\mu + Nn^\mu).(N^\nu + Nn^\nu)$$

To prove the decomposition, it therefore suffices to show that the cross term

$$\tilde{g}^{a0}X^\mu_{,a} + \tilde{g}^{00}N^\mu = 0$$

15.10 The Bianchi identity for non-Abelian gauge fields

Let

$$A = A^a T_a = A^a_\mu dx^\mu T_a$$

be a Lie algebra valued one form. This is the Yang-Mills non-Abelian gauge potential. The corresponding gauge field is

$$F = dA + (1/2)[A, A] = A^a_{\mu,\nu}dx^\nu \wedge dx^\mu T_a + A^a_\mu A^b_\nu dx^\mu \wedge dx^\nu [T_a, T_b]/2$$

and in terms of the structure constants

$$[T_a, T_b] = C(abc)T_c$$

so that

$$F = F^a T_a, F^a = dA^a + C(abc)A^b \wedge A^c$$

Note that

$$A \wedge A = A^a \wedge A^b T_a T_b = -A^b \wedge A^a T_a T_b = -A^a \wedge A^b T_b T_a$$

so

$$[A, A] = A^a \wedge A^b [T_a, T_b] = 2A \wedge A$$

and hence we can equivalently write

$$F = dA + A \wedge A$$

We have

$$dF = d(dA + A \wedge A) = d(A \wedge A) = dA \wedge A + A \wedge dA$$

and hence

$$dF + [A, F] = dA \wedge A + A \wedge dA + A \wedge F - F \wedge A$$

$$= dA \wedge A + A \wedge dA + A \wedge (dA + A \wedge A) - (dA + A \wedge A) \wedge A$$

$$= 0$$

This is the Bianchi identity.

Quantum theory of elasticity: Let $u_k(t, r)$ denote the displacement vector field of an elastic body. Let $F_k(t, r)$ be an externally applied force density field. The Lagrangian density of the body is then

$$L = (1/2)u_{k,t}^2 - (1/2)C(klmn)u_{kl}u_{mn} + F_k(t, r + u(t, r))u_k(t, r)$$

The force field components F_k are assumed to be random fields, say a mixture of a Gaussian and a Poisson field. In this expression, u_{kl} is the strain tensor field

$$u_{kl} = (1/2)(u_{k,l} + u_{l,k})$$

The corresponding momentum density fields are

$$\pi_k = \partial L/\partial u_{k,t} = u_{k,t}$$

and hence the Hamiltonian density is

$$H(u_k, u_{k,m}, \pi_k) = \pi_k u_{k,t} - L = (1/2)\pi_k^2 + (1/2)C(klmn)u_{kl}u_{mn} - F_k(t, r + u)u_k$$

The Schrodinger equation corresponding to this Hamiltonian is then

$$(\int d^3r H(u_k(r), u_{k,m}(r), -i\delta/\delta u_k(r)))\psi(t, u(.)) =$$

$$i\partial_t\psi(t,u(.))$$

If we introduce a small parameter ϵ into the force field $F(t,r)$ by denoting it as $F_\epsilon(t,r)$ having a rate functional $I(F)$, then (a) make a classical computation of the rate function of the resulting solution $u_k(t,r,\epsilon)$ and (b) make a computation of the rate functional of the quantum mechanical wave functional $\psi(t,u,\epsilon)$ and hence calculate for a given observable X defined on the Hilbert space of all square integrable functionals of $u_k(r), r\in D, k=1,2,3$ where D is the spatial domain of the elastic body, the rate functional of the stochastic process $\xi(t,\epsilon) =< \psi(t,.,\epsilon)|X|\psi(t,.,\epsilon)>$, namely the quantum average value of the observable X. Make a comparison between the classical and quantum cases by taking X to be $u_k(.)$.

15.11 More problems in LDP applied to quantum field theory

[1] Let $\phi(x)$ be a field on $\mathbb{R}^n$ with classical field equations given by a Lagrangian $L(\phi,\phi_{,k})$. In the presence of an external current field $J(x)$, the path integral is

$$Z(J)=\int exp(iS(\phi)+i\int J\phi dx)D\phi$$

Let

$$W(J)=log(Z(J))-log\int exp(iS(\phi)+i\int J\phi dx)D\phi$$

where

$$S(\phi)=\int Ldx$$

The average of the quantum field ϕ in the presence of the current field J is given by

$$<\phi>_J(x)=Z(J)^{-1}\int\phi(x)exp(iS(\phi)+i\int J\phi dx)D\phi=-i\delta W(J)/\delta J(x)$$

If $J(x)$ is a random classical current field with given statistics, then what is the statistics of the classical field $<\phi>_J(x)$? Now observe that for a classical field $\chi(x)$, if we define the classical current field J_χ by

$$<\phi>_{J_\chi}=\chi$$

and put

$$\Gamma(\chi)=\int J_\chi(x)\chi(x)dx+iW(J_\chi)$$

then

$$\delta\Gamma(\chi)/\delta\chi(x)$$
$$=J_\chi(x)+\int(\delta J_\chi(y)/\delta\chi(x)))\chi(y)dy+i\int(\delta W(J_\chi)/\delta J(y))(\delta J_\chi(y)/\delta\chi(x))dy$$

$$= J_\chi(x)$$

This is called the quantum equation of motion for the field $\chi(x)$. It is the quantum generalization of the classical equation

$$\delta(S(\phi) + \int J(y)\phi(y)dy)/\delta\phi(x) = J(x)$$

We have an analogue of such a situation in classical probability. Let X be a random variable and $\delta\lambda$ another independent random variable. The moment generating function of X evaluated at $\lambda+\delta\lambda$ taking into account the randomness of $\delta\lambda$ is given by

$$\tilde{M}_X(\lambda) = \mathbb{E}exp((\lambda + \delta\lambda)X)$$

$$= \mathbb{E}(M_X(\lambda + \delta\lambda)) = \int M_X(\lambda + u)f(u)du$$

where f is the probability density of $\delta\lambda$. If $\delta\lambda$ is regarded as weak amplitude noise, then we can write approximately

$$\tilde{M}_X(\lambda) = \sum_{n=0}^{N} M_X^{(r)}(\lambda)\mu_r/r!$$

where

$$\mu_r = \int u^r f(u)du = \mathbb{E}((\delta\lambda)^r)$$

This formula has another kind of significance,namely, while calculating the rate function of a sequence of r.v's of the form S_n/n, $S_n = \sum_{k=1}^n X_k$ where the $X_k's$ are iid, suppose we make an error by using $\lambda+\delta\lambda$ in place of λ in the preliminary step involving the computation of the logarithmic moment generating function. Then the erroneous rate function for this family will be given by

$$\tilde{I}(x) = sup_\lambda(\lambda x - log(\tilde{M}_X(\lambda)))$$

and evaluating this upto the second moment of moment generating function parameter error $\delta\lambda$, we get

$$\tilde{M}_X(\lambda) = M_X(\lambda) + \mu_1.M_X'(\lambda) + \mu_2 M_X''(\lambda)/2$$

$$log(\tilde{M}_X(\lambda)) = log(M_X(\lambda)) + \mu_1 M_X'(\lambda)/M_X(\lambda) + \mu_2 M_X''(\lambda)/2M_X(\lambda)$$
$$-(\mu_1^2/2)(M_X'(\lambda)/M_X(\lambda))^2$$

It follows then that the error in the rate function will be approximately $\delta I(x)$ where

$$I(x) = sup_\lambda(\lambda.x - log(M_X(\lambda))$$

$$I(x)+\delta I(x) = \tilde{I}(x) = sup_\lambda(\lambda x - log(M_X(\lambda)) - \mu_1 M_X'(\lambda)/M_X(\lambda)$$
$$-\mu_2 M_X''(\lambda)/2M_X(\lambda) + (\mu_1^2/2)(M_X'(\lambda)/M_X(\lambda))^2)$$

We leave it as an exercise to evaluate this upto $O(\mu_1, \mu_1^2, \mu_2)$ by showing the additional terms to $I(x)$ that are present.

15.12 Schrodinger and Klein-Gordon equations in quantum field theory based on an infinite dimensional Laplacian operator

Consider the KG Lagrangian for a scalar field ϕ with a nonlinear potential term:

$$L(\phi, \phi_{,\mu}) = (1/2)\partial_\mu\phi.\partial^\mu\phi - m^2\phi^2/2 - V(\phi)$$

The canonical position field is $\phi(x)$ and the canonical momentum field is

$$\pi(x) = \partial L/\partial\partial_0\phi = \partial_0\phi$$

Thus, the Hamiltonian density obtained by applying the Legendre transformation to this Lagrangian is

$$\mathcal{H}(\phi, \nabla\phi, \pi) = \pi.\phi - L = (1/2)(\pi^2 + |\nabla\phi|^2 + m^2\phi^2) + V(\phi)$$

The Hamiltonian is the integral of the Hamiltonian density over the spatial volume:

$$H(\phi, \pi) = \int H(\phi(x), \nabla\phi(x), \pi(x))d^3x$$

$$= \int [(1/2)(\pi(x)^2 + |\nabla\phi(x)|^2 + m^2\phi(x)^2) + V(\phi(x))]d^3x$$

and Schrodinger's equation in infinite dimensions is obtained by assuming that the wave functional $\Psi(t, \phi(r) : r \in \mathbb{R}^3)$ satisfied the differential equation

$$i\frac{\partial\Psi(t, \phi(.))}{\partial t} = H\Psi(t, \phi(.))$$

$$= (1/2)(\int [-\delta^2/\delta\phi(r)^2 + |\nabla\phi(r)|^2 + m^2\phi(r)^2 + 2V(\phi(r))]d^3r)\Psi(t, \phi(.))$$

$$= (1/2)\int [-\delta^2\Psi(t, \phi(.))/\delta\phi(r)^2)+$$

$$|\nabla\phi(r)|^2\Psi(t, \phi(.))+m^2\phi(r)^2\Psi(t, \phi(.))+2V(\phi(r))\Psi(t, \phi(.))]d^3r$$

We can also hope to solve this infinite dimensional Schrodinger equation in which the Hamiltonian appears as an anharmonically perturbed infinite dimensional quantum Harmonic oscillator using the Feymman path integral method. To do this, we choose a cubic box B of side-length L and expand the position field ϕ as a spatial Fourier series

$$\phi(t, r) = \sum_k c(t, k)exp(ik.r)/L^{3/2}, k = (2\pi/L)(k_x, k_y, k_z), k_x, k_y, k_z \in \mathbb{Z}^3$$

Since ϕ is real, we must impose the restrictions

$$c(t, -k) = \bar{c}(t, k)$$

and then we get

$$\int_B L d^3x = T_1 + T_2 + T_3 + T_4$$

where

$$T_1 = (1/2)\int_B (\partial_0\phi(t,r))^2 d^3r = (1/2)\sum_k |\partial_t c(t,k)|^2,$$

$$T_2 = (-1/2)\int_B |\nabla\phi(t,r)|^2 d^3r = (1/2)\sum_k k^2|c(t,k)|^2$$

$$T_3 = (-m^2/2)\int_B \phi(t,r)^2 d^3r = (m^2/2)\sum_k |c(t,k)|^2$$

$$T_4 = -\int_B V(\phi(t,r))d^3r = -\int_B V(\sum_k c(t,k)exp(ik.r)))d^3r$$

$$= -\sum_{n,k_1,...,k_n} V(k_1,...,k_n)c(t,k_1)c(t,k_2)...c(t,k_n)$$

We denote this Lagrangian by

$$L(c(t,k),\partial_t c(t,k) : k \in (2\pi/L)\mathbb{Z}^3$$

and now formulate the corresponding Hamiltonian:

$$\pi(t,k) = \partial L/\partial\partial_t c(t,k)) = (1/2)c(t,-k) = (1/2)\partial_t\bar{c}(t,k)$$

$$\pi(t,-k) = \partial L/\partial\partial_t c(t,-k) = (1/2)\partial_t c(t,k) = \bar{\pi}(t,k)$$

Thus the Hamiltonian in spatial Fourier space is given by

$$H(c(t,k),\pi(t,k) : k \in 2\pi\mathbb{Z}^3/L) =$$

$$\sum_k (\pi(t,k)\partial_t c(t,k) + \pi(t,k)\partial_t\bar{c}(t,k)) - L =$$

$$= (1/2)\sum_k (|\pi(t,k)|^2 + (k^2+m^2)|c(t,k)|^2) +$$

$$+ \sum_{n,k_1,...,k_n} V(k_1,...,k_n)c(t,k_1)c(t,k_2)...c(t,k_n)$$

Schrodinger's equation in terms of this Hamiltonian now appears as

$$i\frac{\partial\Psi(t,c(.))}{\partial t} =$$

$$H(c(k), -i\frac{\partial}{\partial c(k)} : k \in 2\pi\mathbb{Z}^3/L)\Psi(t,c(.))$$

This Schrodinger equation can be solved via a Feynman path integral using an infinite dimensional Brownian motion with a countable number of components with complex time. The unitary evolution kernel is

$$K(T,c|0,d) = C\int_{(0,d)}^{(T,c)} exp(i\int_0^T (1/2)\sum_k |\partial_t c(t,k)|^2 - i\int_0^T W(c(t,.)dt)Dc$$

where

$$W(c(.)) = (1/2)\sum_k ((k^2+m^2)|c(t,k)|^2) + \sum_{n,k_1,...,k_n} V(k_1,...,k_n)c(t,k_1)c(t,k_2)...c(t,k_n)$$

The stationary states of the Harmonic oscillator component are given by

$$|n >= |n_1, n_2, ... >= |n(k) : k \in firsttwoquadrants >= |\{n(k)\} >$$

where

$$c(k)^* c(k)|n >= n(k)|c(k) >$$

The oscillator coherent states are

$$\phi(u) >= \sum_n (\Pi_k u(k)^{n(k)}/\Pi_k \sqrt{n(k)!})|n >$$

Note that

$$|n >= exp(- \| u \|^2 /)\Pi_k \frac{c(k)^{*n(k)}}{\sqrt{n(k)!}}|0 >$$

Then, for k in the first two quadrants

$$c(k)|\phi(u) > -u(k)|phi(u) >, c(k)^* exp(\| u \|^2 /2)|\phi(u) >= \frac{\partial}{\partial u(k)} exp(\| u \|^2 /)\phi(u) >$$

$$= exp(\| u \|^2 /2)(\partial/\partial u(k) + \bar{u}(k)/2))|\phi(u) >$$

or equivalently,

$$c(k)^*|\phi(u) >= (\partial/\partial u(k) + \bar{u}(k)/2)|\phi(u) >$$

Chapter 16

Weak Convergence, Sanov's Theorem, LDP in Binary Signal Detection

16.1 Prohorov's tightness theorem, "necessity part"

Let X be a separable metric space and $M(X)$ a set of probability measures on the Borel subsets of X. Then if $M(X)$ is compact only if for each $\epsilon > 0$, there exists a compact $K \subset X$ such that $\mu(K) > 1 - \epsilon \forall \mu \in M(X)$. To prove this, first observe that since X is separable, for each $n = 1, 2, ...$, we have

$$X = \bigcup_{r=1}^{\omega} S(n, r)$$

where $S(n, r)$ is an open ball of radius $1/n$. We now show that for each n and each $\delta > 0$ there exists a finite positive integer $k(n)$ such that for all $\mu \in M(X)$, we have that

$$\mu(\bigcup_{r=1}^{k(n)} S(n, r)) > 1 - \delta$$

Indeed if this is not the case, then there is a $\delta > 0$ and a finite positive integer $N = N(n, \delta)$ such that for each $m > N$ there is a $\mu_m \in M(X)$ such that

$$\mu_m(\bigcup_{r=1}^{m} S(n, r)) \leq 1 - \delta$$

Since $M(X)$ is compact, there is an increasing sequence $m(l)$ of positive integers such that $\mu_{m(n)}$ converges weakly to a measure $\mu \in M(X)$. Thus

$$\mu(\bigcup_{r=1}^{m(k)} S(n,r)) \leq liminf_l \mu_{m(l)}(\bigcup_{r=1}^{m(k)} S(n,r))$$

$$\leq liminf_l \mu_{m(l)}(\bigcup_{r=1}^{m(l)} S(n,r)) \leq 1-\delta$$

and then letting $k \to \infty$ we get a contradiction:

$$\mu(X) \leq 1-\delta$$

Now given an $\epsilon > 0$, in accordance with the above result, for each n, choose a positive integer $k(n)$ so that

$$\mu(\bigcup_{r=1}^{k(n)} S(n,r)) > 1-\epsilon/2^n \forall \mu \in M(X)$$

Then, it follows from the union bound that

$$\mu(\bigcap_n \bigcup_{r=1}^{k(n)} S(n,r)) \geq 1-\epsilon \forall \mu \in M(X)$$

and noting that

$$K = \bigcap_n \bigcup_{r=1}^{k(n)} \bar{S}(n,r)$$

is compact, the proof is complete. Note that

$$S(n,r) = \{x : d(x, x(n,r)) < 1/n\}, \bar{S}(n,r) = \{x : d(x, x(n,r)) \leq 1/n\}$$

for some $x(n,r) \in X$. In fact, we can choose a countable dense set $\{x(n) : n \geq 1\}$ in X and then define

$$S(n,r) = \{x : d(x, x(r)) < 1/n\}$$

16.2 Sanov's theorem on the LDP for empirical distributions of discrete iid random variables

Let μ be the probability distribution that assigns probability $\mu(i)$ to i where $i = 1, 2, ..., n$. Thus, $\mu(i) > 0$ and $\sum_{i=1}^{N} \mu(i) = 1$. Let L_N denote the empirical probability distribution on $\{1, 2, ..., n\}$ that is obtained from N independent r.v's $X_1, ..., X_N$, each one having the distribution μ. Thus,

$$L_N = N^{-1}.\sum_{i=1}^{N} \delta_{X_i}$$

It is clear that for each $i = 1, 2, ..., N$, $L_N(i)$ assumes only the values k/N where $k = 0, 1, ..., N$. Let $q(i), i = 1, 2, ..., n$ be such that for each i, $Nq(i)$ is a non-negative integer and $\sum_{i=1}^{n} q(i) = 1$. Then,

$$P(L_N(i) = q(i), i = 1, 2, ..., N) = \frac{N!}{\Pi_{i=1}^{n}((Nq(i))!)} \Pi_{i=1}^{n} \mu(i)^{Nq(i)}$$

by the multinomial distribution theorem. Indeed, this is the probability that in N independent simulations of a r.v. having distribution μ, i will occur $Nq(i)$ times for each $i = 1, 2, ..., N$. Taking logarithms and using the Stirling approximation, we get

$$N^{-1}.log(P(L_N(i) = q(i), i = 1, 2, ..., N)) \approx$$

$$N^{-1}(N + 1/2)log(N) - N^{-1} \sum_{i=1}^{n} (Nq(i) + 1/2)log(Nq(i)) + \sum_{i=1}^{N} q(i)log(\mu(i))$$

This approximation is valid in the sense that the difference between the two sides is $o(1/N)$ as $N \to \infty$. Letting $N \to \infty$ then gives us the celebrated theorem of Sanov:

$$lim_{N \to \infty} N^{-1}.log(P(L_N(i) = q(i), i = 1, 2, ..., N))$$

$$= -\sum_{i=1}^{N} q(i)log(q(i)/\mu(i)) = -H(q|\mu)$$

namely, the negative of the relative entropy between the probability distributions q andμ. From this result we easily deduce that if E is a set of probability distributions on $\{1, 2, ..., n\}$, then

$$P(L_N \in E) = \sum_{q \in E} P(L_N = q) = \sum_{q \in E} exp(-NH(q|\mu) + o(1))$$

and hence

$$P(L_N \in E)^{1/N} = [\sum_{q \in E} exp(-NH(q|\mu) + o(1))]^{1/N}$$

$$\to sup_{q \in E} exp(-H(q|\mu))$$

or equivalently,

$$N^{-1}.log(P(L_N \in E)) \to -inf_{q \in E} H(q|\mu)$$

which is another form of the celebrated theorem of Sanov.

16.3 LDP in binary phase shift keying

Under H_1, the measured signal is $x(t) = s_1(t) + \sqrt{\epsilon}w(t)$ and under H_0, the measured signal is $x(t) = s_0(t) + \sqrt{\epsilon}w(t)$ where $t \in [0, T]$ and $w(t)$ is white Gaussian noise with spectrum σ^2. The optimum likelihood ratio test decides that H_1 is correct if

$$\int_0^T s(t)x(t)dt > \eta$$

where

$$s(t) = s_1(t) - s_0(t), \eta = (E_1 - E_0)/2, E_k = \int_0^T s_k(t)^2 dt, k = 0, 1$$

and decides H_0 otherwise. The error probabilities are therefore

$$P(H_1|H_0) = P(\int_0^T s(t)x(t)dt > \eta|H_0) =$$

$$P(\sqrt{\epsilon}\int_0^T s(t)w(t)dt > (E_1 + E_0)/2 - R)$$

where

$$R = \int_0^T s_1(t)s_0(t)dt$$

and

$$P(H_0|H_1) = P(\int_0^T s(t)x(t)dt < \eta|H_1) =$$

$$P(\sqrt{\epsilon}\int_0^T s(t)w(t)dt < -(E_1 + E_0)/2 + R))$$

where

$$R = \int_0^T s_1(t)s_0(t)dt$$

By symmetry of the normal distribution, both of these error probabilities are equal. Denoting the error probabiity by $P_\epsilon(e)$, and setting

$$A = (E_1 + E_0)/2 - R$$

we the find that

$$lim_{\epsilon \to 0}\epsilon.log(P_\epsilon(e)) =$$

$$lim_{\epsilon \to 0}\epsilon.log(P(\sqrt{\epsilon}\int_0^T s(t)w(t)dt > A)$$

$$= -\delta^2/2$$

where

$$\delta = A/\sigma \parallel s \parallel = A/\sigma\sqrt{E_1 + E_2 - 2R} = \sqrt{A/2\sigma^2}$$

What if $w(t)$ is non-Gaussian white noise with a given rate function ? In this case, we can write $w(t) = v'(t)$ where $v(.)$ is a Levy process and thus has a moment generating functional of the form

$$\mathbb{E}exp(\int_0^T f(t)dv(t)) = exp(\int_0^T \Lambda(f(t))dt)$$

If we assume that $v(t)$ is the limit as $N \to \infty$ of a process of the form $v_N(t) = N^{-1}\sum_{n=1}^{[Nt]} X(n)$ where the $X(n)'s$ are iid zero mean random variables with logarithmic moment generating function

$$\Lambda(\lambda) = log\mathbb{E}[exp(\lambda X_(1))]$$

then the family of processes $v_N(.), N \geq 1$ converges to the zero process with a rate functional of

$$sup_f(\int_0^T f(t)x'(t)dt - \int_0^T \Lambda(f(t))dt)$$

$$= \int_0^T sup_z(zx'(t) - \Lambda(z))dt = \int_0^T \Lambda^*(x'(t))dt$$

where

$$\Lambda^*(u) = sup(zu - \Lambda(z))$$

For weak noise, the likelihood ratio reduces to

$$P(x|H_1)/P(x|H_0) \approx exp(-(1/\epsilon)(I_w(x-s_1))/exp(-(1/\epsilon)I_w(x-s_0))$$

and hence in the weak noise limit $\epsilon \to 0$, the test becomes: Choose H_1 if $I_w(x-s_0) > I_w(x-s_1)$ and choose H_0 otherwise. The error probability is then

$$P(H_0)P(I_w(x-s_0) > I_w(x-s_1)|H_0) + P(H_1)P(I_w(x-s_1) > I_w(x-s_0)|H_1)$$

$$= P(H_0)P(I_w(\sqrt{\epsilon}w) > I_w(s_0 - s_1 + \sqrt{\epsilon}w))$$

$$+P(H_1)P(I_w(\sqrt{\epsilon}w) > I_w(s_1 - s_0 + \sqrt{\epsilon}w))$$

16.4 Compactness of the set of probability measures on a compact metric space

Let X be a compact metric space and let $M(X)$ denote the set of all probability measures on X. Then , $M(X)$ is compact. To see this we first observe that $C(X)$ is separable since X is compact by the Stone-Weierstrass theorem. Thus, we can choose a countable dense subset $\{f_n : n \geq 1\}$ in $C(X)$. Since X is compact, each $f \in C(X)$ is bounded. Thus, $\| f_n \| < \infty \forall n$. We write f_n for $f_n / \| f_n \|$. Then $\| f_n \| = 1 \forall n$. For any $\mu \in M(X)$, define $T(\mu) = \{\int f_n d\mu : n \geq 1\}$. Then $T(\mu) \in I^\infty$ where $I = [0,1]$. We have seen that $T(M(X))$ is closed. In fact, suppose $\mu_k \in M(X)$ and $T(\mu_k)$ converges in I^∞. Then write

$$\alpha = \{\alpha(n)\} = lim_k T(\mu_k)$$

We have to show the existence of a $\mu \in M(X)$ such that $T(\mu) = \alpha$. We wish to show that μ_k converges weakly to a measure (probability) μ. To do so, we choose any $f \in C(X)$. Choose an n and a real number c such that $\| f - cf_n \| < \epsilon$ where ϵ is any given positive number. Then by the triangle inequality,

$$|\int f d\mu_k - \int f d\mu_l| \leq 2 \| f - cf_n \| + |c| |\int f_n d\mu_k - \int f_n d\mu_l|$$

which converges to a number smaller than 2ϵ as $k, l \to \infty$. Hence $\int f d\mu_k$ converges to some $\Lambda(f)$. We easily verify that $\Lambda(.)$ is a linear positive functional on $C(X)$ and that $\Lambda(1) = 1$. Thus, by a version of the Riesz representation theorem, there is a probability measure μ on X such that

$$\Lambda(f) = \int f d\mu, \forall f \in C(X)$$

In other words,

$$lim \int f d\mu_k = \int f d\mu \forall f \in C(X)$$

establishing the weak convergence of μ_k to μ. In particular,

$$\alpha = lim T(\mu_k) = T(\mu)$$

proving that $T(M(X))$ is closed. T is also continuous since if $\mu_k \to \mu$, then

$$\int f_n d\mu_k \to \int f_n d\mu \forall n$$

Further T is invertible on its range $T(M(X))$ because suppose $T(\mu) = T(\nu)$ for some $\mu, \nu \in M(X)$. Then

$$\int f_n d\mu = \int f_n d\nu \forall n$$

and hence if f is any element of $C(X)$, then for any given $\epsilon > 0$, we choose n so that $\| f - cf_n \| < \epsilon$ and then

$$|\int f d\mu - \int f d\nu| \leq 2 \| f - cf_n \| + |c| |\int f_n d\mu - \int f_n d\nu|$$

$$< 2\epsilon$$

so $\epsilon > 0$ being arbitrary, we get

$$\int f d\mu = \int f d\nu \forall f \in C(X)$$

and therefore $\mu = \nu$. Further, T^{-1} is continuous on $T(M(X))$ since if $T(\mu_n) \rightarrow T(\mu)$, then for any $f \in C(X)$, we choose n such that $\| f - cf_n \| < \epsilon$ and then

$$|\int f d\mu_k - \int f d\mu| \leq 2 \| f - cf_n \| + 2|c|| \int f_n d\mu_k - \int f_n d\mu|$$

which converges to something smaller that 2ϵ as $k \rightarrow \infty$ proving thereby that $\mu_k \rightarrow \mu$. Thus, we have proved that

$$T : M(X) \rightarrow T(M(X) \subset [0, 1]$$

is a homeomorphism. with $T(M(X))$ being closed and hence compact in $[0, 1]$. Thus, $M(X)$ is also compact.

16.5 Large deviations for frequency modulated signals

Let $s(t)$ be a message signal and before applying it to a phase modulator, it gets corrupted by WGN $\sqrt{\epsilon}w(t)$. Let $B(t) = \int_0^t w(s)ds$. Then $B(.)$ is standard Brownian motion and if $m(t) = \int_0^t s(\tau)d\tau$, the FM signal is given by

$$x(t) = cos(\omega t + Km(t) + K\sqrt{\epsilon}B(t))$$

To demodulate this signal, we differentiate this signal and calculate the envelope of the resulting signal. This envelope is given by

$$\omega + Ks(t) + K\sqrt{\epsilon}w(t)$$

and we can analyze this demodulated FM signal using the standard methods of LDP. Now suppose to transmit the bits 1 or 0, we use such a phase modulation method. For transmitting 1, we transmit $cos(\omega t)$ and to transmit 0, we transmit $-cos(\omega t)$, ie we change the phase by π. When noise corrupts the phase, we tramsmit $x(t) = cos(\omega t + K\sqrt{\epsilon}B(t)$ and to transmit 0, we transmit $x(t) = cos(\omega t + \pi + K\sqrt{\epsilon}B(t) = -cos(\omega t + \sqrt{\epsilon}B(t))$. The standard method of decoding involves correlating the received signal with $cos(\omega t)$ over the time interval $[0, T]$ where $T = 2\pi/\omega$ and if the resulting correlation is closer to $T/2$ than to $-T/2$, we decide that a one was sent and otherwise decide that a zero was sent. Then, the error probabilities are

$$P(0|1) = P(1|0) = P(\int_0^T cos(\omega t).cos(\omega t + K\sqrt{\epsilon}B(t))dt > 0)$$

We leave it as an exercise to apply LDP theory to calculate an approximate expression for this probability as $\epsilon \to 0$ in the form $exp(-I_0/\epsilon)$ where I_0 is the infimum of the rate function $I(f) = \int_0^T f'(t)^2/2$ of Brownian motion over the set

$$E = \{f : \int_0^T cos(\omega t).cos(\omega t + Kf(t))dt > 0\}$$

Note that

$$E = \{f : (1/2)\int_0^T cos(Kf(t))dt + (1/2)\int_0^T cos(2\omega t + Kf(t))dt > 0\}$$

Reference:

A.Dembo and O.Zeitouni, "Large deviations, Techniques and Applications", Springer.

Chapter 17

LDP for Classical and Quantum Transmission Lines, String Theoretic Corrections to Classical Field Lagrangians, Non-Abelian Gauge Theory in The Language of Differential Forms

17.1 LDP theory applied to transmission lines with line loading

The line equations taking into account voltage and current loading by a white noise process, ie the differential of an independent increment process which will have a rate function of the form

$$I_T(f) = \int_0^T \Lambda^*(f(t))dt$$

are

$$\partial_z v(t,z) + Ri(t,z) + L\partial_t i(t,z) = W_v(t,z)$$

$$\partial_z i(t,z) + Gv(t,z) + C\partial_t v(t,z) = W_i(t,z)$$

where W_v, W_i are white Gaussian in time, so that they admit representations

$$W_v(t,z) = \sum_n a_n(z)B'_n(t), W_i(t,z) = \sum_n b_n(z)B'_n(t)$$

with the $B'_n s$ being independent Brownian motion processes. We expand $v(t,z)$ and $i(t,z)$ as Fourier series in the z variable:

$$i(t,z) = \sum_{n\in\mathbb{Z}} i_n(t)exp(2\pi inz/d),$$

$$v(t,z) = \sum_{n\in\mathbb{Z}} v_n(t)exp(2\pi inz/d)$$

to get the following Ito stochastic differential form of the above line equations:

$$L.di_n(t) + Ri_n(t)dt + (2\pi in/d)v_n(t)dt = \sum_m a(n,m)dB_m(t),$$

$$Cdv_n(t) + Gv_n(t)dt + (2\pi in/d)i_n(t)dt = \sum_m b(n,m)dB_m(t)$$

where

$$a(n,m) = d^{-1}\int_0^d a_n(z)exp(-2\pi imz/d)dz, b(n,m)$$

$$= d^{-1}\int_0^d b_n(z)exp(-2\pi imz/d)dz$$

These equations can be expressed in matrix-vector notation as

$$d\begin{pmatrix} i_n(t) \\ v_n(t) \end{pmatrix} =$$

$$\begin{pmatrix} -R/L & (-2\pi in/dL) \\ (-2\pi in/dC) & -G/C \end{pmatrix}\begin{pmatrix} i_n(t) \\ v_n(t) \end{pmatrix} dt + \sum_m \begin{pmatrix} a(n,m)/L \\ b(n,m)/C \end{pmatrix} dB_m(t)$$

In order to obtain the rate function for the processes i_n, v_n over the time interval $[0,T]$ in the case when the noise is weak so that the $B'_n s$ are replaced by $\sqrt{\epsilon}B_n$ with $\epsilon \to 0$, we shall require to carry out the following optimization: minimize $\int_0^T f(t)^2 dt$ subject to the constraint $dx(t)/dt = Ax(t) + bf(t), x \in \mathbb{R}^2, b \in \mathbb{R}^2, A \in \mathbb{R}^{2\times 2}$. In this case it is a trivial optimization problem for the differential equation implies the constraint

$$b_2(x'_1 - (Ax)_1) = b_1(x'_2 - (Ax)_2)$$

Assuming both b_1, b_2 are non-zero, it then follows that the rate functional of the x process can be expressed as

$$I(x) = (1/2b_1^2)\int_0^T (x'_1(t) - (Ax)_1(t))^2 dt = (1/2b_2^2)\int_0^T (x'_2(t) - (Ax)_2(t))^2 dt$$

17.2 LDP formulation of the quantum transmission line

The line equations can be expressed as

$$x'(t) = Ax(t) + GW(t), W = B'$$

where x is an infinite dimensional vector and B is an infinite dimensional standard Brownian motion vector. This system of linear stochastic differential equations can be derived from Heisenberg matrix mechanics using a Hamiltonian for harmonic oscillators along with Lindblad operators. Specifically, $x(t)$ is a Markov diffusion process with generator

$$L = -x^T A\nabla_x + (1/2)Tr(GG^T\nabla\nabla^T)$$

Given any function $f(x)$ on the state space, we compute its differential using the Ito rule

$$df(x) = f'(x)bdB + Lf(x)dt$$

This suggests a quantum realization of this the process using the Evans-Hudson flow:

$$dj_t(f) = j_t(\partial_k f)G_{kj}dB_j + j_t((1/2)\partial_k\partial_m f)H_{km}dt$$

for $j_t(f) = f(x(t))$. In the quantum context, we replace f by an observable X, $(1/2)H_{km}\partial_k\partial_m$ by a linear map θ_0 on the space of observables, $G_{kj}\partial_k$ by another linear map θ_j on the space of observables and B_k by $A_k + A_k^*$ where $A_k(t)$ and $A_k(t)^*$ are respectively the annihilation and creation processes in the quantum stochastic calculus of Hudson and Parthasarathy. Then, the above commutative Evans-Hudson flow becomes the non-commutative Evans-Hudson flow

$$dj_t(X) = j_t(\theta_0(X))dt + j_t(\theta_k(X))(dA_k + dA_k^*)$$

In general, given an general Evans-Hudson flow

$$dj_t(X) = j_t(\theta_b^a(X))d\Lambda_a^b(t)$$

where $\Lambda_b^a(t), a, b = 0, 1, 2, ...$ satisfying the quantum Ito formula

$$d\Lambda_b^a(t).d\Lambda_d^c(t) = \epsilon_b^c d\Lambda_d^a(t)$$

we can raise the question that when θ_b^a for $(a, b) \neq (0, 0$ are scaled by a small parameter ϵ, then for a given time t, what is the rate functional of the family of probability distributions of $j_t(X) = j_t(X, \epsilon)$ as $\epsilon \to 0$ in a given state of the system and bath, say the bath is in a coherent state ? More generally, what is the rate function of the time averaged observable $\int_0^T j_t(X)dt$?

Another formulation is to write down the differential equations for the spatial Fourier series components of the line current and voltage and then derive these differential equations using the Heisenberg-Lindblad equations for open quantum systems with the Hamiltonian being that of an infinite sequence of harmonic oscillators and the Lindblad operators being linear functions of the creation and annihilation operators of the harmonic oscillator Hamiltonian. Thus we obtain a quantum open system theoretic formulation of the transmission line and we can derive this master equation from the Hudson-Parthasarathy noisy Schrodinger (HPS) equation unitary dynamics by tracing out over the bath. Instead of tracing out over the bath, we retain the HPS equation but introduce small perturbation parameters into the Lindblad operators and then attempt to derive after Belavkin filtering of the state of the system and bath a rate functional for the evolving filtered density operator (The Belavkin filter is a classical stochastic Schrodinger equation since the measurement noise algebra is Abelian.

17.3 Yang-Mills gauge fields, The Euler characteristic, string theoretic corrections to the Yang-Mills anomaly cancellation Lagrangian terms

Consider for a positive integer m,

$$\chi_m = Tr(F^m) = Tr(F \wedge ... \wedge F)$$

m-fold wedge product where

$$F = dA + A \wedge A$$

For example, when $m = 1$, we have

$$\chi_1 = Tr(F \wedge F)$$

$$F = F^a T_a$$

$$d\chi_2 = Tr(dF \wedge F + F \wedge dF) = (dF^a \wedge F^b)Tr(T_a T_b) + (F^b \wedge dF^a)Tr(T_b T_a)$$
$$= dF^a \wedge F^b.Tr(\{T_a, T_b\})/2$$

because F^a is a two form and hence $dF^a \wedge F^b = F^b \wedge dF^a$. Likewise,

$$\chi_2 = Tr(F \wedge F \wedge \wedge F \wedge F)$$

$$d\chi_2 = Tr(dF \wedge F \wedge F \wedge F) + Tr(F \wedge dF \wedge F \wedge F)$$
$$+Tr(F \wedge F \wedge dF \wedge F) + Tr(F \wedge F \wedge F \wedge dF)$$
$$= (dF^a \wedge F^b \wedge F^c \wedge F^d)Tr(T_a T_b T_c T_d)$$
$$+(F^b \wedge dF^a \wedge F^c \wedge F^d)Tr(T_b T_a T_c T_d)$$

$$+(F^b \wedge F^c \wedge dF^a \wedge F^d)Tr(T_bT_cT_aT_d)$$
$$+(F^b \wedge F^c \wedge F^d \wedge dF^a)Tr(T_bT_cT_dT_a)$$
$$= (dF^a \wedge F^b \wedge F^c \wedge F^d)(Tr(\{T_a, T_b\}T_cT_d)/2 + Tr(T_bT_c\{T_a, T_d\}/2))$$

Consider now

$$\Omega_1 = d(a,b)F^a \wedge F^b$$

where

$$d(a,b) = d(b,a)$$

We then have

$$d\Omega_1 = d(a,b)dF^a \wedge F^b + d(a,b)F^a \wedge dF^b$$

Using the Bianchi identity

$$dF^a + C(abc)A^b \wedge F^c = 0$$

gives us

$$d\Omega_1 = -d(a,b)C(acd)A^c \wedge F^d \wedge F^b - d(a,b)C(bcd)F^a \wedge A^c \wedge F^d$$
$$= -(d(a,b)C(acd) + d(b,a)C(acd))A^c \wedge F^b \wedge F^d$$
$$= -2d(a,b)C(acd)A^c \wedge F^b \wedge F^d$$

Using the fact that $A^c \wedge F^b \wedge F^d$ is symmetric w.r.t interchange of b and d and we can show that this equals zero. For example, choosing $d(a,b) = \delta(a,b)$, we have to show that $C(acd)A^c \wedge F^a \wedge F^d = 0$. But this follows from the antisymmetry of the structure constants combined with the symmetry of $A^c \wedge F^a \wedge F^d$ w.r.t. interchange of a and d.

Note: The Jacobi identity gives

$$[T_a, [T_b, T_c]] + [T_b, [T_c, T_a]] + [T_c, [T_a, T_b]] = 0$$

or equivalently,

$$[T_a, C(dbc)T_d] + [T_b, C(dca)T_d] + [T_c, C(dab)T_d] = 0$$

or equivalently,

$$C(dbc)C(ead) + C(dca)C(ebd) + C(dab)C(ecd) = 0$$

More generally, we can show using the Bianchi identity that if $d(a_1, ..., a_n)$ is a totally symmetric tensor, ie symmetric and invariant under the Lie algebra, then

$$\Omega = d(a_1, ..., a_n)F^{a_1} \wedge ... \wedge F^{a_n}$$

is a closed form, ie

$$d\Omega = 0$$

This computation is based on

$$d\Omega = d(a_1, ..., a_n)(dF^{a_1} \wedge F^{a_2} \wedge ... \wedge F^{a_n}$$

$$+(F^{a_1} \wedge dF^{a_2} \wedge F^{a_3} \wedge ... \wedge F^{a_n} + ..+$$

$$F^{a_1} \wedge ... \wedge F^{a_{n-1}} \wedge dF^{a_n})$$

$$= -d(a_1, ..., a_n)(C(a_1bc)A^b \wedge F^c \wedge F^{a_2} \wedge ... \wedge F^{a_n}+$$

$$+C(a_2bc)F^{a_1} \wedge A^b \wedge F^c \wedge F^{a_3} \wedge ... \wedge F^{a_n}+$$

$$+... + C(a_nbc)F^{a_1} \wedge ... \wedge F^{a_{n-1}} \wedge A^b \wedge F^c)$$

Yang-Mills Chern-Simons form: Let

$$\omega_Y = Tr(A \wedge F + 2A \wedge A \wedge A/3)$$

where

$$F = dA + 2A \wedge A$$

Note that the Yang-Mills covariant differential is

$$D = d + A$$

so that

$$F = [D, D] = [d + A, d + A] = dA + [A, A] = dA + 2A \wedge A$$

Then,

$$d\omega_Y = Tr(F \wedge F)$$

Proof:

$$d(A \wedge F) = dA \wedge F - A \wedge dF$$

The Bianchi identity is

$$dF = d(A \wedge A) = dA \wedge A - A \wedge dA = (F - 2A \wedge A) \wedge A - A \wedge (F - 2A \wedge A)$$

$$= [F, A] = -[A, F]$$

so that

$$dF + [A, F] = 0$$

which is the Bianchi identity. Thus,

$$d(A \wedge F) = dA \wedge F - A \wedge dF = dA \wedge F + A \wedge [A, F]$$

$$= dA \wedge F + A \wedge A \wedge F - A \wedge F \wedge A$$

$$= (F - 2A \wedge A) \wedge F + A \wedge A \wedge F - A \wedge F \wedge A$$

$$= F \wedge F - A \wedge A \wedge F - A \wedge F \wedge A$$

Further,

$$d(A \wedge A \wedge A) = dA \wedge A \wedge A - A \wedge dA \wedge A + A \wedge A \wedge dA$$

$$= (F - 2A \wedge A) \wedge A \wedge A - A \wedge (F - 2A \wedge A) \wedge A + A \wedge A \wedge (F - 2A \wedge A)$$

$$= F \wedge A \wedge A - A \wedge F \wedge A + A \wedge A \wedge F$$

$$-2A \wedge A \wedge A \wedge A$$

Then, let

$$\omega = A \wedge F + g.A \wedge \wedge A \wedge A$$

We then find that

$$d\omega =$$

$$F \wedge F - A \wedge A \wedge F - A \wedge F \wedge A$$

$$+g.F \wedge A \wedge A - gA \wedge F \wedge A + gA \wedge \wedge A \wedge F - 2gA \wedge A \wedge A \wedge A$$

Now

$$Tr(A \wedge A \wedge A \wedge A) = (A^a \wedge A^b \wedge A^c \wedge A^d)Tr(T_a T_b T_c T_d) - - - (1)$$

Using the properties of the trace and the total antisymmetry of $A^a \wedge A^b \wedge A^c \wedge A^d$ w.r.t all its four indices, we can show that (1) evaluates to zero.

It is known that if F is the Yang-Mills curvature and R the spinor represon tation of the gravitational curvature, then there exist three forms ω_Y and ω_L such that locally

$$d\omega_Y = Tr(F \wedge F), d\omega_L = Tr(R \wedge R)$$

In particular locally,

$$d(\omega_Y - \omega_L) = Tr(F \wedge F) - Tr(R \wedge R)$$

From supergavity theory, it is known in fact that globally also there exists a 3-form H such that

$$dH = Tr(F \wedge F) - Tr(R \wedge R)$$

This is called the Bianchi identity with string theoretic corrections. In fact, as shown in the volumes on superstring theory by Green, Schwarz and Witten, the presence of a counterterm in the supergravity action involving H or equivalently $\omega_Y - \omega_L$ is required in order to cancel out the gravitation and gauge anomalies of the supersymmetric action arising from Hexagon diagrams. The term involving ω_Y is anyway present in the original low energy supersymmetric action but the second term ω_L is a purely string theoretic correction which breaks the supersymmetry but ensures anomaly cancellations and can indeed be shown to be the lowest order string theoretic correction to the low energy action.

Estimating using the EM algorithm pulse parameters. $p(t|\phi)$ is a pulse depending upon non-random parameters ϕ . The received signal is

$$x(t) = \sum_{k=1}^{n} A(k)p(t - t_k - \delta t_l|\phi) + \sqrt{\epsilon}w(t)$$

where $\{\delta t_l\}$ is a random vector with a pdf $f(\delta t|\eta)$, η being another set of parameters. The total parameter vector to be estimated is $\theta = (\phi, \eta)$ from the measurements $x(t), t \in [0, T]$. This is estimated using the EM algorithm treating δt as the latent variable. The EM algorithm maximizes

$$Q(\theta, \theta_0) = \int log(p(x, \delta t|\theta))p(\delta t|x, \theta_0)d\delta t$$

or equivalently,

$$Q_1(\theta, \theta_0) = \mathbb{E}[log(p(x|\delta t, \phi)) + log(f(\delta t|\eta)))|x, \theta_0]$$

or equivalently,

$$Q_2(\theta, \theta_0) = \int (log(p(x|\delta t, \phi)) + log(f(\delta t|\eta)))p(x|\delta t, \phi_0)f(\delta t|\eta_0)d\delta t$$

17.4 Quantum averaging based derivation of action functional for point fields from action functional of string fields

A heuristic approach to deriving low energy effective approximate action for point function fields based on exact high energy action functionals of string fields.

Consider a quantum string field

$$X^\mu(\sigma) = x^\mu + \delta X^\mu(\sigma)$$

where $\delta X^\mu(\sigma)$ is a quantum field built out of Bosonic and Fermionic creation and annihilation operators. Let ρ be the state of the quantum system and ϕ a point field. Consider a Lagrangian density $L(\phi(x), \phi_{,\mu}(x))$ of this point field. The quantum averaged string theory corrected action functional is

$$S[\phi] =< \int L(\phi(x + \delta X(\sigma)), \phi_{,\mu}(x + \delta X(\sigma)))d\sigma d^4x >$$

where for any observable X, $< X >$ denotes $Tr(\rho X)$. This action functional is quasi classical. A purely quantum string theoretic quantum effective action functional based on the Feynman path integral for fields would be

$$S_q[\phi] = -i.log(\int exp(i \int L(\phi(x+\delta X(\sigma)), \phi_{,\mu}(x+\delta X(\sigma))+S_0[\delta X]))d^4x d\sigma)D\delta X)$$

where $S_0[\delta X]$ is the action functional for the string perturbation field $\delta X(\sigma)$. This action functional is simply the sum of the bosonic and Fermionic action functionals for the string (See Green, Schwarz and Witten, "Superstring Theory").

Let $f(x|\phi)$ be a signal field on $\mathbb{R}^n$ with ϕ an unknown parameter vector on which the field depends. The measured field is a filtered version of this field:

$$g(x) = \int L(x,y|\eta)f(y|\phi)dy + W(x)$$

where η is a random parameter vector having pdf $p(\eta|\chi)$ dependent upon some other unknown parameter vector χ. The total parameter vector is $\theta = (\phi, \chi)$. $W(x)$ is a spatially white Gaussian noise field and hence the probability density functional of $g(x), x \in D$ is given by

$$p(g|\theta) = C. \int exp((-1/2N_0)) \int_D (g(x) - \int L(x,y|\eta)f(y|\phi)dy)^2 dx).p(\eta|\chi)d\eta$$

The maximum likelihood estimate of θ given the measurements $g(x), x \in D$ can be obtained by maximizing this functional. The complexity of this algorithm is too high. The EM algorithm considerably simplifies this to the iteration

$$\theta_1 = argmax_\theta Q(\theta, \theta_0)$$

where

$$Q(\theta, \theta_0) = \int log(p(g,\eta|\theta))p(\eta|g,\theta_0)d\eta$$
$$= \mathbb{E}(log(p(g,\eta|\theta))|g,\theta_0)$$

or equivalently,

$$-2N_0 Q(\theta, \theta_0) = \int_D \mathbb{E}((g(x) - \int L(x,y|\eta)f(y|\phi)dy)^2|g,\theta_0)dx$$
$$2N_0 \mathbb{E}(log(p(\eta|\chi))|g,\theta_0)$$

Noting that

$$p(\eta|g,\theta_0) = p(g|\eta,\phi_0)p(\eta|\chi_0)/p(g|\theta_0)$$

the EM algorithm is equivalent to minimizing

$$Q_1(\theta, \theta_0) = Q_1(\phi, \chi, \phi_0, \chi_0) =$$
$$(\int_D (g(x) - \int L(x,y|\eta)f(y|\phi)dy)^2 dx)p(g|\eta,\phi_0)p(\eta|\chi_0)d\eta$$
$$-2N_0 \int log(p(\eta|\chi)).p(g|\eta,\phi_0).p(\eta|\chi_0)d\eta$$

Chapter 17

LDP and EM Algorithm, LDP for Parameter Estimates in Linear Dynamical Systems, Philosophical Questions in Quantum General Relativity

18.1 Large deviations and the EM algorithm: Large deviation properties of parameter estimates derived using the EM algorithm in the presence of noise relative to ML parameter estimates obtained in the absence of noise when there are latent random parameter vectors in the measurement model

Let $\theta(n,x)$ denote the nth iteration of the estimate of a parameter vector θ based on the EM algorithm. Assume that the measured signal x has the form $x(t) = s(t,\eta|\theta) + w(t,\epsilon)$ where θ is the parameter to be estimated. η is a random vector independent of w having a pdf $p(\eta|\theta)$. The EM algorithm estimates θ using the recursion

$$\theta(n+1,x) = argmax_\theta \mathbb{E}[log(p_{w_\epsilon}(x - s(.,\eta|\theta)).p(\eta|\theta))|x,\theta_0]$$

This conditional expectation may be replaced by multiplication with $p(x|\eta,\theta_0)p(\eta|\theta_0)$ followed by integration w.r.t η. Note that the exact mle of θ is obtained by maximizing

$$\int p(x|\eta,\theta)p(\eta|\theta)d\eta =$$

$$\int p_{w_\epsilon}(x - s(.,\eta|\theta))p(\eta|\theta)d\eta$$

Now suppose noise is absent, ie, $\epsilon = 0$. Then the mle of θ would be obtained by maximizing

$$\int p(x|\eta,\theta)p(\eta|\theta)d\eta$$

$$= \int \delta(x - s(.,\eta|\theta))p(\eta|\theta)d\eta$$

The LDP problem is now immediate to formulate: As $\epsilon \to 0$ at what rate does the difference between the estimate of θ with noise $w(.,\epsilon)$ and the estimate without noise, ie, $\epsilon = 0$ converge to zero ? Assume in both the cases that the random parameter η is the same with known pdf $p(\eta|\theta)$.

18.2 Fundamental problems in quantum general relativity

[1] The problem of time. There is only one origin of time, namely the big-bang. Entropy of the universe must keep increasing since the big bang and this is incompatible with the unitary evolution of the state of the universe since unitary dynamics is reversible and in fact preserves the Von-Neumann entropy.

[2] The problem of measurement of the system which is the universe. According to the Copenhagen interpretation of quantum mechanics, a measurement should cause the state to collapse and further evolution then starts from the collapsed state. However, the measurement apparatus must necessarily be within our universe and hence we cannot talk of the measuring apparatus as being outside our system. Hence, state collapse following a measurement must indeed be a part of the unitary dynamics of the system.

[3] The states of a macroscopic body must be represented as a very large number of tensor products of the states of its constituent particles and hence the inner product between any two such microscopic states must vanish because such an inner product is a product of a very large number of terms, each being smaller than unity in magnitude. This means that when a state is a superposition of several pure states with each component state being represented as a tensor product of a very large number of pure states then while forming the norm square of the resultant state, cross terms, ie, interference terms will be absent and hence quantum effects will not appear, ie, the resulting state will exhibit classical behaviour. Likewise when we take successive measurements of a mixed

state taking into account the collapse postulate after each measurement, the sum of all the resulting probabilities will be unity as interference terms will be negligible. More precisely, if ρ is the mixed state and $\{P_k\}$ is a projection valued measurement taken at times $t_1 < t_2 < ... < t_N$ with the unitary evolution U_k taking place in the time interval (t_{k-1}, t_k), then the joint probability of $(m_1, ..., m_N)$ occurring is given by

$$P(m_N, ..., m_1) = Tr(P(m_N)U_N P(m_{N-1})U_{N-1}...P(m_1)U_1\rho U_1^* P(m_1)...U_{N-1}^* P(m_{N-1})U_N^* P(m_N))$$

and this will sum up over $m_1, ..., m_N$ to unity because the cross terms like

$$P(m_N, ..., m_1; k_N, ..., k_1)$$
$$= Tr(P(m_N)U_N P(m_{N-1})U_{N-1}...P(m_1)U_1\rho U_1^* P(k_1)...U_{N-1}^* P(k_{N-1})U_N^* P(k_N))$$

with

$$(k_N, ..., k_1) \neq (m_N, ..., m_1)$$

will be negligible.

Large deviation problems: Time is an observable in background independent quantum field theory as can be seen from the following situation: Let H, T be observables. We wish to model T as a time observable that evolves under the unitary dynamics generated by H according to

$$T(t) = U(t)^* TU(t), U(t) = exp(-itH)$$

We now take another observable X and wish to define its value at a time T_0 where T_0 is another observable. To do so, we must define it to be $X(t) = U(t)^* XU(t)$ where t is an operator valued time index that satisfies $T(t) = T_0$, ie $X(T_0)$ is defined to be X at that time t when the time observable T assumes the operator value T_0. If T_0 is a small perturbation of a scalar times the identity, then we are nearly in the classical situation with a small quantum non-commutative perturbation and we can raise questions about the large deviation properties of $X(T_0)$ in a given state. In general, suppose $|\lambda>$ is an eigenstate of T_0 with the eigenvalue λ. Then we can define a time $t = t(\lambda)$ at which $T(t)|\lambda> = T_0|\lambda> = \lambda|\lambda>$.If T commutes with T_0 so that $|\lambda, s>$ is a joint eigenstate of T_0 with eigenvalue λ and of T with eigenvalue s. Then on the state $|\lambda, s>$ we can define X at time T_0 to be $X(s) = U(s)^* XU(s)$ and the time observable T assumes the value λ on this very same state.

18.3 A problem in Large deviations and Lie algebras

When the parameters of a linear dynamical system undergo small random fluctuations with a given rate function, then determine the rate function of the state process of the dynamical system using Lie algebraic representations of the exponential function of the sum of two matrices.

Let

$$A(\theta) = A_0 + \sum_{k=1}^{p} \theta(k) A_k$$

with the $A_k's$ being matrices and the $\theta(k)'s$ small random parameters. Assume that $\theta(k) = \theta(k, \epsilon)$, ie, as a vector $\theta = \theta(\epsilon)$ with $\theta(\epsilon)\epsilon \to 0$ satisfying an LDP with rate $I(\theta)$. Then compute the rate function of the process

$$x(t, \epsilon) = \int_0^t \Phi(t - s|\theta(\epsilon))u(s)ds$$

where $u(t)$ is another random process independent of the $\theta(\epsilon)$. Note that x solves the differential equation

$$dx(t)/dt = A(\theta)x(t) + u(t)$$

hint: Writing

$$exp(tA(\theta)) = exp(tA_0).G(t)$$

we derive

$$G'(t) = \sum_{k=1}^{p} \theta(k) exp(-tad(A_0))(A_k)G(t)$$

and hence

$$G(t) = I + \sum_{n\geq 1, k_1, ..., k_n} \theta(k_1)...\theta(k_n) \int_{0<t_n<...<t_1<t} exp(-t_1 ad(A_0))(A_{k_1})...exp(-t_k ad(A_0))(A_{k_n})dt_1...dt_n$$

18.4 Describing quantum gravity using holonomy fields

Let $A(x)$ be a $\mathfrak{g}$ valued connection field where $\mathfrak{g}$ is the Lie algebra of the gauge group $G \subset SU(N)$. The parallel displacement $H(t) \in G$ of $H(0)$ along a curve $t \to x(t)$ using this connection is given by

$$\frac{dH(t)}{dt} = H(t)A(x(t))$$

which formally integrates to give

$$H(t) = \Phi(t)H(0)$$

where

$$\Phi(t) = I + \sum_{n\geq 1} \int_{0<t_n<...<t_1<t} A(x(t_n))...A(x(t_1))dt_1...dt_n$$

If however we wish to take space-time coordinates into account while describing parallel displacement along a space-time curve $s \rightarrow x^\mu(s)$ using the Lie algebra connection $A_\mu(x)dx^\mu = A^a_\mu(x)\tau_a.dx^\mu$, then the formula for the parallel displacement/gauge group holomomy is given by

$$\frac{dH(s)}{ds} = H(s)A_\mu(x(s))\frac{dx^\mu(s)}{ds}$$

or equivalently

$$dH(s) = H(s)A_\mu(x(s))dx^\mu(s)$$

While quantizing the Yang-Mills field, we can use as our position fields either $A_\mu(x)$ or instead adopt loop quantization in which we discretize the space-time manifold using a graph of curves γ. For example, we choose a 3-dimensional surface Σ and draw curves from different discretely chosen points on this surface. Then we replace the position field $A_\mu(x)$ by the Lorentz group value fields defined on the space of these curves γ. Let $\gamma(b)$ denote the beginning point of the curve γ and $\gamma(f)$ the final point. Then, $H(\gamma)$ is define as $H(1)$ where $\gamma(b) = \gamma(0), \gamma(f) = \gamma(1)$ and the parallel displacement gives

$$\frac{dH(s)}{ds} = H(s)A_\mu(\gamma(s))\frac{d\gamma^\mu(s)}{ds}, 0 \leq s \leq 1$$

with $H(0)$ being the identity matrix.

Consider now the situation of the gravitational field. The gravitational connection can be described in the spinor representation as

$$\Gamma_\mu(x) = \Gamma^\mu_{ab}(x)[\gamma^a, \gamma^b]/4$$

where γ^a are the Dirac matrices. Note that $[\gamma^a, \gamma^b]/4, 0 \leq a < b < 3$ form a standard set of Lie algebra generators for the four dimensional Lorentz group. The spinor representation of the Riemann gravitational curvature tensor is

$$R_{\mu\nu} = R_{\mu\nu ab}[\gamma^a, \gamma^b]/4 = [\Gamma_\mu, \Gamma_\nu]$$

If we look at the ADM formalism for canonical quantum gravity, then we are imbedding a three dimensional manifold Σ_t at each time t in four dimensional space-time and then the position fields become the spatial components $q_{ab}, 1 \leq a, b \leq 3$ of the metric tensor. The gauge group is then $SO(3)$ rather than the Lorentz group $SO(3, 1)$.

18.5 Square of the Dirac operator in curved space-time in the presence of a non-Abelian Yang-Mills gauge potential

Consider the Dirac operator in curved space-time and in the presence of a non-Abelian gauge field.

$$D = \gamma^\mu \nabla_\mu, \nabla_\mu = D_\mu + A_\mu, D_\mu = \partial_\mu + \Gamma_\mu$$

We have

$$D^2 = \gamma^\mu \nabla_\mu \gamma^\nu \nabla^\nu$$
$$= \gamma^\mu([\nabla_\mu, \gamma^\nu] + \gamma^\nu \nabla_\mu)\nabla_\nu$$
$$= \gamma^\mu \gamma^\nu \nabla_\mu \nabla_\nu$$

because

$$[\nabla_\mu, \gamma^\nu] = 0$$

because the gravitational covariant derivative of the tetrad vanishes. Note that $\gamma^\mu = \gamma^a e_a^\mu(x)$ where e_a^μ is the gravitational tetrad. Thus

$$D^2 = ((1/2)\{\gamma^\mu, \gamma^\nu\} + (1/2)[\gamma^\mu, \gamma^\nu])\nabla_\mu \nabla_\nu$$
$$= g^{\mu\nu}\nabla_\mu \nabla_\nu + (1/4)[\gamma^\mu, \gamma^\nu][\nabla_\mu, \nabla_\nu]$$

Now observe that

$$[\nabla_\mu, \nabla_\nu] = R_{\mu\nu} + F_{\mu\nu}$$

where $R_{\mu\nu}$ is the gravitational curvature in the spinor representation while $F_{\mu\nu}$ is the Yang-Mills curvature. So we can write

$$D^2 = \Delta + R + F$$

where Δ is the Laplace-Beltrami operator of the gravitational and Yang-Mills connection while R and F are respectively the gravitational curvature and Yang-Mills curvature fields in the spinor representation:

$$R = (1/4)[\gamma^\mu, \gamma^\nu]R_{\mu\nu}, F = (1/4)[\gamma^\mu, \gamma^\nu]F_{\mu\nu}$$

18.6 Exponential equivalence, The Dawson-Gartner theorem on LDP on projective limits with applications to estabilishing LDP for processes or more generally, LDP on the projective limit of topological spaces

18.7 Applications

[a] Schilder's theorem for the rate function of Brownian motion.

[b] Rate function for the Poisson process.

Suppose

18.8 Equivalence of LDP

:

Let $Z(\epsilon), Y(\epsilon), \epsilon \to 0$ be two families of random variables defined on the same probability space with values in the same Banach space. Suppose that for all $\delta > 0$,

$$lim_{\epsilon \to 0}\epsilon.log(P(\| Z(\epsilon) - Y(\epsilon) \| > \delta)) = -\infty$$

Then, if $Y(\epsilon)$ satisfies an LDP with rate function I, $Z(\epsilon)$ will also satisfy an LDP with the same rate function I. In fact, this follows from

$$P(Z(\epsilon) \in E) \leq P(Z(\epsilon) \in E, \| Z(\epsilon) - Y(\epsilon) \| \leq \delta) + P(\| Z(\epsilon - Y(\epsilon) \| > \delta)$$

so

$$limsup_{\epsilon \to 0}\epsilon.log(P(Z(\epsilon) \in E) \leq limsup_{\epsilon \to 0}\epsilon.log(P(Y(\epsilon) \in E(\delta)))$$

because

$$lim_{\epsilon \to 0}P(\| Z(\epsilon) - Y(\epsilon) \| > \delta) = -\infty$$

Here,

$$E(\delta) = \{x : d(x, E) \leq \delta\}$$

where

$$d(x, E) = \inf\{\| x - y \| : y \in E\}$$

From the LDP upper bound for Y we deduce that

$$limsup_{\epsilon \to 0}P(Z(\epsilon) \in E) \leq -inf\{I(y) : y \in \bar{E}(\delta)\}$$

Letting $\delta \to 0$ then gives the desired upper bound for the LDP of Z. The lower bound follows by a similar argument with Y and Z interchanged. Specifically,

$$P(Y(\epsilon) \in E) \leq P(Z(\epsilon) \in E(\delta)) + P(\| Z(\epsilon) - Y(\epsilon) \| > \delta)$$

from which using the lower bound for Y, we get

$$-inf\{I(y) : y \in E(\delta)^o\} \leq lim_{\epsilon \to 0}\epsilon.log(P(Z(\epsilon) \in E(\delta))$$

and letting $\delta \to 0$, the desired result follows.

LDP for projective limits: Let $X_k, k \in I$ be a family of sets with a partial order $<$ on I. If $k < l$, we assume that there is a map (called projection) $p_{kl} : X_l \to X_k$ so that $p_{mk}op_{kl}$ for $m < k < l$. We define X to be the set of all points $x = (x(k) : k \in I)$ in $\Pi_{k \in I}X_k$ for which $p_{kl}(x(l)) = x(k)$ for all $k < l$. Define $p_k : X \to X_k$ so that $p(x) = x(k) \forall k \in I$. Then X is called the projective limit of X. Note that $p_{kl}op_l = p_k$ on X for all $k < l$. Now suppose that we are given a family of probability measures $\mu_\epsilon, \epsilon \to 0$ on X such that for each $k \in I$, $\mu_\epsilon op_k^{-1}$ satisfies an LDP in X_k with rate function I_k. Then we claim that if these rate functions are good rate functions, ie, for each finite real number a and each $k \in I$, $\psi_k(a) = I_k^{-1}((-\infty, a])$ is compact, then μ_ϵ satisfies an LDP in X with good rate function $I(x) = sup\{I_k(p_k(x)) : k \in I\}$.

18.9 Lehmann's representation of the propagator of the Klein-Gordon field with nonlinear perturbations

The free, ie, unperturbed KG field satisfies

$$(\Box + m^2)\phi(x) = 0$$

and hence if the free propagator is defined as

$$D(x-y) =< 0|T(\phi(x)\phi(y))|0 >$$

then

$$(\Box + m^2)D(x) = \delta^4(x)$$

In deriving this, we use the equal time CCR

$$[\phi(t,r), \partial_t\phi(t,r')] = i\delta^3(r-r')$$

which is obtained from the expression for the canonical momentum field derived from the KG Lagrangian. Thus, the free KG propagator in the four momentum domain is

$$D(p) = (p^2 - m^2 - i0)^{-1}, p^2 = (p^0)^2 - P^2$$

Equivalently, in the space-time domain, using contour integration for the p^0 integral, we obtain by closing the contour via the upper half plane infinite semicircle that

$$D(x) = \int D(p)exp(ip.x)d^4p = \int (2E(P))^{-1}.exp(ip.x)|_{p^0=E(P)}d^3P$$

for $x^0 > 0$ and for $x^0 < 0$, we have by closing the contour via the lower half plane infinite semicircle that

$$D(x) = \int (2E(P))^{-1}exp(ip.x)|_{p^0=-E(P)}d^3P = D(x,m^2)$$

where

$$E(P) = \sqrt{m^2+P^2} = E(P,m)$$

Note that

$$\int \delta(p^2-m^2)\theta(p^0)2p^0exp(ip.x)d^4p$$

$$= \int (2E(P))^{-1}exp(ip.x)|_{p^0=E(P)}d^3P$$

because of the identity

$$\delta(p^2-m^2) = (2E(P))^{-1}(\delta(p^0-E(P)) - \delta(p^2+E(P)))$$

In general, when there are perturbing nonilearities in the KG Lagrangian, we can write

$$< 0|\phi(x)\phi(y)|0 >=$$

$$= \sum_{n\geq 1} \int < 0|\phi(x)|p_1, ..., p_n >< p_1, ..., p_n|\phi(y)|0 > d^3p_1...d^3p_n$$

Now,

$$\int < 0|\phi(x)|p_1, ..., p_n >< p_1, ..., p_n|\phi(y)|0 > d^3p_1...d^3p_n$$

$$= \int | < 0|\phi(0)|p_1, ..., p_n > |^2 exp(-i(p_1 + ... + p_n).(x-y))|_{p_j^0=E(P_j)} d^3P_1...d^3P_n$$

$$= \int \Pi_{j=1}^n (\delta(p_j^2 - m^2)2p_j^0)\theta(p_j^0))exp(-iq.(x-y))\delta(q-p_1-...-p_n)d^4p_1...d^4p_nd^4q$$

From the Lorentz invariance of this expression, we can express it as

$$\int \rho(q^2)exp(-iq.(x-y))d^4q$$

where

$$\rho(q^2) = \int < 0|\phi(0)|p_1, ..., p_n > \Pi_{j=1}^n(\delta(p_j^2-m^2)(2p_j^0\theta(p_j^0))\delta(q-p_1-...-p_n)d^4p_1...d^4p_n$$

Note that $\rho(q^2)$ is concentrated on the region $q^0 > 0$ because of the constraint functions $\theta(p_j^0), j = 1, 2, ..., n$ and the function $\delta(q - p_1 - .. - p_n)$ appearing in the above integral. Then,

$$\int \rho(q^2)\theta(q^0)exp(-iq.(x-y))d^4q = \int \delta(q^2-m^2)\theta(q^0)\rho(m^2)exp(-iq.(x-y))d(m^2)d^4q$$

$$= \int \rho(m^2)d(m^2) \int (2(E(Q,m))^{-1}exp(-iq.(x-y))d^3Q$$

$$= \int \rho(m^2)D(x-y, m^2)d(m^2)$$

where

$$E(Q, m) = \sqrt{Q^2 + m^2}$$

Chapters Index

Chapter 1:LDP problems in quantum field theory

[1.1] Large deviations for supergravity fields. LDP for supergravity Lagrangians perturbed by small random supersymmetry breaking Lagrangians.

[1.2] Rate function of string propagator. Change in bosonic string propagator and quantum string amplitudes caused by a small perturbing random gauge field interacting with the string.

[1.3] Large deviations for p-form fields. Generalizations of the electromagnetic field to higher dimensions based on p-form Lagrangians perturbed by random terms. LDP rate function for the perturbed field.

[1.4] The dynamics of the electroweak theory. Electroweak matter and gauge fields perturbed by random classical current and random classical gauge potential fields:Study of this perturbation on the rate function of quantum average values of observables defined as functionals of the gauge and matter fields.

[1.5] Quantum filtering in Fermionic noise: The Fermionic Belavkin quantum filter based on the Hudson-Parthasarathy bosonic and fermionic stochastic calculus. Calculating the classical stochastic differential equations satisfied by the filtered observable and the filtered state, Rate function for the filtered observable and state when the Lindblad noise parameters in the HP QSDE become very small.

[1.6] Quantum field theory is a low energy limit of string field theory. String theoretic corrections to the action for the gravitational and gauge fields, evaluation of the propagator for the Klein-Gordon quantum field in the presence of external random quantum potential fields with and LDP analysis of the quantum mechanical scattering amplitudes when the perturbing potential is small.

[1.7] LDP problems associated with the Atiyah-Singer index theorem. If the gauge potential in the Yang-Mills-Dirac operator in curved space-time has a small random component, then what will be the effect of random perturbations in the gauge potential on the index of this Dirac operator ? Since the index is an integer, one can ask the question, how large should the gauge potential perturbation be so that the index becomes a non-degenerate (integer valued) random variable ?

[1.8] LDP problems in general relativity.

[1] Perturbation to the geodesic trajectory of a charged particle in curved space-time metric caused by a small random electromagnetic field.

[2] Perturbation to the string field caused by the presence of a gauge field; perturbation to the interaction Lagrangian between the gauge field and the string field caused by small quantum fluctuations in the string field as well by perturbations to the string field caused by the gauge field.

Chapter 2:LDP in biology, neural networks, electromagnetic measurements, cosmic expansion

[2.1] The importance of mathematical models in medicine.

Mathematical models in biology. The living body is modeled as a black box whose outputs like blood velocity/density field, speech signals, EEG signals etc. are described by stochastic differential/partial differential equations with unknown parameters and input signals given by stimuli. The unknown parameters characterize the nature of the disease and some of these parameters may also be known and controllable. The idea is to estimate the unknown parameters from noisy output measurements and hence determine the probability that these unknown parameter estimates differ from nominal ones by more than a threshold using the large deviation principle with weak noise assumption. This probability is expressed in terms of the control parameters which are then modified using appropriate medicine so that the probability is minimized.

[2.2] LDP related problems in neural netwrorks and artificial intelligence. A system is modeled as a stochastic differential equation with unknown parameters and weak noise. A neural network is trained to output the parameter estimates with input given by the output of the sde. The training is done by varying the parameters of the sde, generating its output and feeding this output to the input of the nn and matching its output with the parameter values. For a given choice of parameters, the probability that it will differ from the neural network output by an amount more than a threshold is computed using LDP theory. This procedure assesses th performance of the nn.

[2.3] LDP problems in cosmic expansion and general relativity. The linearized Einstein field equations are driven by the energy-momentum tensor of the weak cosmic microwave background electromagnetic radiation field which is assumed to be a Gaussian field. The problem is then to calculate the rate functional of the metric, velocity and density perturbations and use this rate function to determine the probability of a rare spike which can cause the evolving inhomogeneities like galaxies to get deformed in unexpected ways.

[2.4] LDP problems in biology: When the permittivity and permeability fields of a tissue undergo small random fluctuations, with known statistics, then we estimate the non-random components of these fields from measurements of the scattered electromagnetic radiation when an incident em field falls on the tissue. We calculate using LDP the probability that these parameter estimates will deviate from nominal values by amounts more than a threshold, this is the probability that a disease has set in, we then control some of the parameters of the tissue using medicine so that this deviation probability is minimized.

[2.5] A Sensitive Quantum Mechanical Method for Measuring the scattered Electromagnetic Fields. A non-random em field is incident upon a Dirac electron bound to its nucleus. When it is incident, a small random component of the field with known statistics is introduced and we wish to reject its effect on the atomic transitions. We introduced control parameters into the non-random field components and calculate using time dependent perturbation theory and LDP the shift in the transition probability due to the random field component. This shift is a random quantity and we use the LDP to calculate the probability that this shift in the transition probability is small, ie, exceeds a threshold with minimum probability. This deviation probability is minimized by adjusting the controllable parameters of the non-random field components.

Chapter 3:LDP in signal processing, communication and antenna design

[3.1] Review

[3.2] LDP problems in dual-band ssb modulation. Given two families of random processes with known rate functions, we have to derive the LDP rate function for a linear combination of the corresponding lowpass envelopes. This LDP will enable to assess the effect of noise on modulation systems involving transmission of lowpass signals over a channel by transforming them into band-pass signals so that these signals are compatible with the channel characteristics.

Large deviation rate function for the Hilbert transform of a family of random processes. The basic idea is that suppose $x_\epsilon(t)$ is a family of random processes having rate functional $I(x)$ and if these processes are transmitted through a channel having impulse response $h(t)$, then the rate function of the family of output random processes $y_\epsilon(t) = h(t)*x_\epsilon(t)$ can be calculated by the contraction principle of Dawson and Gartner or equivalently, by assuming that this transfer function is invertible so that if $g(t)$ is the inverse impulse response of $h(t)$, then the rate function of the y family is $I_1(y) = I(g*y)$ and if it is not invertible, then the rate function is $I_1(y) = inf(I(x), x : h*x = y)$

[3.3] What is meant by estimating a quantum field in space-time ?

[a] Solutions to some problems in antenna design with LDP applications: In any antenna design process, say involving the design of an array or a current distribution so as to match the generated radiation pattern to a given pattern, if there are small random fluctuations in the given pattern, then on solving the matching/optimization problem, there will correspondingly be small fluctuations in the sensor array positions and their currents or in the shape and surface current density of the designed antenna. Then, what is the probability in terms of the statistics of the fluctuations of the given pattern of the radiation from the designed antenna deviating by a large amount from the desired noiseless pattern ?

[b] Estimating a quantum field in space-time using the Belavkin filter.

[3.4] Noisy Schrodinger equation for and N-particle system and derivation of the one particle quantum Boltzmann equation in the presence of noise by partial tracing over the remaining particles.

[3.5] The pde's satisfied by the quantum electromagnetic field observables in a cavity.

[3.6] Estimating the quantum state of a single particle system in a system of N indistinguishable particles.

Chapter 4: LDP applied to quantum measurement, classical Markov chains, quantum stochastics and quantum transition probabilities

[4.1] Some other aspects of measurement of a quantum field. The quantum field is specified by a family of operators (quantum field operators) in Fock space, say a tensor product of a Boson and a Fermionic Fock space. This family of operators is indexed by a spatial position variable and these operators evolve with time according to a field Hamiltonian in accordance with the Heisenberg matrix mechanics picture. The state of the field at time $t = 0$ is specified. In the Schrodinger picture, this state evolves according to the dual of the Heisenberg dynamics under the same Hamiltonian, ie, according to the adjoint action of the unitary evolution operator. In this Schrodinger picture, the field operators do not evolve with time. At time t_1 we measure the Schrodinger evolved state using the eigenbasis of the field operator at position r_1 and accordingly apply the state collapse postulate. After this collapse, the state evolves under the Schrodinger dynamics to another state at time t_2 and then we measure the field operator at position r_2 using its eigenbasis and then again after noting the measurement outcome, apply the state collapse postulate. In this way, we sequentially take measurements of the field operators at positions $r_1, ..., r_N$ at times $t_1, ..., t_N$ so that following every measurement, state collapse occurs and in the time durations between any two successive measurements, free Schrodinger evolution occurs. The LDP problem here is: If there are small fluctuations in the field operators as well as in the initial state of the system, then how will these fluctuations reflect upon the joint probability distribution of the above sequential measurements as well as upon the final state of the system ? If these fluctuations are viewed as operator valued classical random variables parametrized by a small parameter, then what will be the probability distribution of the final state as well as of the joint probabilities of the sequential measurements ? Can one obtain a rate function for these as the parameter converges to zero in terms of the classical statistics of these operator fluctuations ?

[4.2] Problems and solutions in antenna theory.

[4.3] Some additional LDP related problems in classical and quantum antenna theory.

[4.4] LDP for quantum Markov chains using discrete time quantum stochastic flows.

[4.5] **LDP applied to the analysis of the error process in stochastic filtering theory of a continuous time Markov process when the measurement noise is white Gaussian and more generally when the measurement noise is the differential of a Levy process (ie, a limit of compound Poisson process plus white Gaussian noise).**

[4.6] LDP applied to the electroweak theory

[4.7] LDP problems in quantum mechanical transitions

[4.8] LDP for the probability distribution of an observable in a quantum Gaussian state perturbed by a small anharmonic potential.

[4.9] Large deviation problems in queueing theory

[4.10] Large deviations in quantum stochastic process theory.

[a] The process is the sum of a quantum Brownian motion $A(t) + A(t)^*$ and a quantum Poisson process $\Lambda(t)$. The quantum Ito formulas are $dA.dA^* = dt, dA^*.dA = 0, d\Lambda(t).dA(t)^* = dA(t)^*, dA(t).d\Lambda(t) = dA(t), (d\Lambda(t))^2 = d\Lambda(t)$. More generally, for a vector $m \in \mathcal{H}$ and a self-adjoint operator H in $\mathcal{H}$ commuting with the time spectral measure we can define quantum processes $A_t(m), A_t(m)^*$ and $\Lambda_H(t)$ satisfying the quantum Ito formulas $dA_t(m).dA_t(m)^*d =<< m, m >> (t)$, $(d\Lambda_H(t))^2 = d\Lambda_{H^2}(t), d\Lambda_H(t).dA_t(m)^* = dA_t(Hm)^*$ and $dA_t(m).d\Lambda_H(t) = dA_t(Hm)$. The problem is to calculate the rate function of the single observable $X = \int_0^T f(t)d\Lambda_H(t) + \int_0^T g(t)(dA_t(m) + dA_t(m)^*)$ in the coherent state $|\phi(u) >$ by evaluating its moment generating function and then its Legendre transform. Note that this single observable will have a rate function which will tell us at what rate $\epsilon.X \to 0$ as $\epsilon \to 0$ but the process $X(t) = \int_0^t f(s)d\Lambda_H(s) + \int_0^t g(s)(dA_s(m) + dA_s(m)^*), 0 \leq t \leq T$ will not have a rate function because it is non-commutative and hence we cannot speak of a joint probability distribution of the paths of this process in any state.

[b] In the Belavkin filter based on non-demolition measurement processes, the state is estimated dynamically and this filtered state satisfies a stochastic Schrodinger equation and hence this being a commutative equation, we can determine the rate functional of the filtered state process using Schilder's formula for the rate function of the driving Wiener process coming from the measurement process. This is the quantum analogue of finding the rate functional of the conditional probability density of the state at time t given measurements upto time t using the Kushner-Kallianpur filtering equations.

Chapter 5: LDP in classical stochastic process theory and quantum mechanical transitions

[5.1] **Large deviations problems to the propagation of noise at the sigmoidal computation nodes through the neural network**

[31],[32],[33]

[5.2] Law of the iterated logarithm for sums of iid random variables.

[5.3] A version of the LDP for iid random variables.

[5.4] The law of the iterated logarithm for sums of iid random variables.

[5.5] An open problem relating applications of LDP to martingales.

[5.6] Properties of ML estimators based on iid measurements.

[5.7] Large deviations for hypothesis testing based on iid measurements:rate function for the error probability under the Neyman-Pearson criterion.

[5.8] Large deviations in quantum cosmology. Approximate the ADM action and Hamiltonian of space-time as the sum of a harmonic oscillator Hamiltonian and small cubic terms under canonically chosen position and momentum fields in terms of the metric tensor. Calculate the transition probabilities induced by the cubic terms between two stationary states of the harmonic oscillator Hamiltonian. Calculate the rate at which these transition probabilities converge to zero in the limit as the cubic terms become very small.

[5.9] Test on Stochastic processes and queueing theory. Calculating the equilibrium distribution of the number of customers in the queue, the equilibrium distribution of the waiting time using ladder variables, model a queue using Markov chain theory and the Chapman-Kolmogorov equations. When the Markov chain infinitesimal generators get perturbed by small random amounts, then determine the LDP rate function for the state probabilities obtained by solving the Chapman-Kolmogorov equation.

Chapter 6: LDP in pattern recognition and Fermionic quantum filtering

[6.1] Large deviation problems in pattern recognition. Derivation of the sigmoidal function representation for the likelihood ratio with specialization to the case of Gaussian distributions. Asymptotic statistical properties of the sigmoidal function based on LDP rate function analysis.

[6.2] LDP for estimating the parameters in mixture models. The distribution of the observation given the class to which a latent variable belongs is known. The probability of the latent variable belonging to a given class is also known. Given the latent variable, the parameters on which the distribution of the observation depend are known. The aim is to estimate these parameters using the maximum likelihood method applied to the observation alone, ie, after integrating out over the latent variables. In the limit when the number of independent observations becomes very large the asymptotics of the parameter estimates are to be calculated, ie, the rate at which these parameter estimates will converge to the true parameter.

[6.3] The EM algorithm: The distribution of the observation given the parameters is usually a very complex function for optimization over the parameters. However, given latent random variables, the distribution of the observation will usually have a very elementary dependence upon the parameters. The EM algorithm exploits this fact to derive an easily implementable recursive algorithm for parameter estimation that converges to the ML estimates. Determining the rate of convergence of the EM parameter estimates to the true ML parameter estimates is the basic LDP problem.

[6.4] Sanov's theorem and Gibbs distributions. The empirical distribution of a sequence of iid random variables obeys a large deviation principle with the rate given according to Sanov's theorem by the relative entropy function. In statistical mechanics, we are interested in choosing that distribution which the asymptotic empirical distribution attains with maximum probability conditioned on an energy constraint on the empirical distribution. The solution to this problem is obtained according Sanov's theorem by minimizing the relative entropy function subject to the energy constraint. This leads to all the well known distributions of statistical mechanics.

[6.5] Gibbs distribution for interacting particle systems. Here, again we wish to choose that distribution which the empirical distribution for iid random variables assumes with maximum probability subject to quadratic energy constraints on the empirical distribution. The quadratic constraint corresponds to the energy of mutually interacting particles in contrast the previously used linear constraint which corresponds to the individual energy of the particles caused by an external field. In general, in the presence of external fields and mutual interactions, we get the sum of linear and quadratic functionals of the empirical distribution given which, we have to maximize the probability that the empirical distribution will assume a given value. In the asymptotic limit, the optimal distribution is according to Sanov's theorem, that distribution which minimizes the relative entropy subject to the linear-quadratic constraint on the same distribution. The relative entropy is calculate w.r.t the true distribution of the random variables.

[6.6] Inversion of the characteristic function.

[6.7] Preliminary result for proof of the Levy-Khintchine formula for infinitely divisible distributions.

[6.8] Existence of a stationary distribution for an irreducible aperiodic Markov chain.

[6.9] Lecture on quantum filtering in the presence of Fermionic noise

[6.10] Lecture plan for Pattern Recognition

[6.11] Review of the book Stochastics, control and robotics,

[6.12] Chapters in "A survey of advanced quantum mechanics and related topics"

Chapter 7: LDP in spin field theory, anharmonic perturbations of quantum oscillators, small perturbations of quantum Gibbs states

[7.1] **Large deviation problems in spin-field interaction theory**:

[7.2] Large deviation problems associated with a quantum gravitational field interacting with a non-Abelian gauge field. A quantum gravitational field interacts with a non-Abelian Yang-Mills gauge field. We formulate the Hamiltonian integral for this total field in which the position fields are the spatial components the metric tensor and the constraint functions appearing in the form of Lagrange multipliers and the Yang-Mills gauge potentials. The Hamiltonian is a highly nonlinear function of the position and momentum fields and on discretizing the spatial manifold into pixels we get an approximation to this Hamiltonian. We can calculate the Gibbs density of this Hamiltonian and study the asymptotic behaviour of the expected value of a field observable in this Gibbs state in the asymptotic limit as the number of spatial pixels goes to infinity.

[7.3] Large deviation problems in quantum harmonic oscillator problems with nonlinear terms. Calculate the transition probabilities regarding the nonlinear terms as small time independent perturbations to the harmonic oscillator Hamiltonian, then in the presence of a small random electromagnetic field interacting with the charge on the harmonic oscillator, calculate using time dependent perturbation theory the approximate transition probabilities and in the limit as this random em field converges to zero, determine the rate at which the transition probabilities converge to the same in the absence of radiation but with the anharmonic terms present.

[7.4] Limiting logarithmic moment generating functional and its Legendre transform for quantum stochastic processes in a given mixed state.

[7.5] Large deviations in the equilibrium distribution of a Markov chain when the generator contains a small random parameter.

Chapter 8: LDP for electromagnetic control of gravitational waves, randomly perturbed quantum fields, Hartree-Fock approximation, renewal processes in quantum mechanics

[8.1] **Gravitational wave propagating in a background curved space-time, LDP for reducing the wave fluctuations via electromagnetic control**

[8.2] The Lehmann representation of the propagator: $<0|\phi(x)\phi(y)|0>$ is to be evaluated using a sum over n-particle states with $n = 1, 2, ...$

$$<0|\phi(x)\phi(y)|0>=\sum_n \int <0|\phi(x)|p_1,...,p_n><p_1,...,p_n|\phi(y)|0> dp_1...dp_n$$

$$=\sum_n \int |<0|\phi(0)|p_1,...,p_n>|^2 exp(i(p_1+...+p_n,x-y))dp_1...dp_n$$

$$=\int exp(i(P,x-y))dP\sum_n \int |<0|\phi(0)|p_1,...,p_n>|^2\delta(P-p_1-...-p_n)dp_1...dp_n$$

[8.3] The LDP problem in the context of a small random non-linearly perturbed Klein-Gordon field:Computation of the statistics of the propagator.

[8.4] The central limit theorem for renewal processes.

[8.5] Applications of renewal process theory in quantum field theory: Calculation of the transition probabilities for Hamiltonians defined as functionals of position and momentum fields taking into account the interaction Hamiltonian between the fields and random current sources generated by electrons hitting a detector plate at renewal times.

[8.6] Large deviation principle in the Hartree-Fock method for approximately solving many electron problems. Evaluation of the rate function of the quantum average of an observable given by a function of the electron positions and momenta when a small random electromagnetic field interacts with the electrons. Evaluation of the asymptotic quantum average of an observable as a function of the number of electrons when the number of electrons becomes very large.

[8.7] LDP in fuzzy neural networks and in quantum neural networks. When there are small random fluctuations in the input and desired output, then what is the LDP rate functional of the neural weight process ?

[8.8] Large deviations analysis of this quantum neural networks.

[8.9] LDP problems in quantum field theory related to corrections to the electron, photon and non-Abelian gauge boson propagators. The effect of non-Abelian gauge fields and the gravitational field on the electron's mass.

Chapter 9: LDP in electromagnetic scattering and string theory, control of dynamical systems using LDP

[9.1] A summary of a list of LDP applications in physics and engineering

[9.2] LDP problems related to scattering of electromagnetic waves by a perfectly conducting cylinder in a curved background space-time.

[9.3] String theory and large deviations:Evaluation of the change in the action functional of a point particle field when the point particle is replaced by a quantum string, ie its position is specified by that of a classical point particle plus a quantum string fluctuation with a length parameter expressible as a superposition of Boson and Fermion creation and annihilation operators.

[9.4] Questions related to qualitative properties of quantum noise

[9.5] Questions on Pattern Recognition

[9.6] Appendix on designing control parameters to minimize the deviation probability of a dynamical system from the stability zone, and on questions related to constructing Lie group invariants.

String theoretic corrections to classical field theory Lagrangians

Chapter 10: LDP in Markov chain and queueing theory with quantum mechanical applications

[10.1] Notes on applications of LDP to stochastic processes and queueing theory

[10.2] LDP problems in Markov chain theory

[10.3] Continuity and non-differentiability of the Brownian sample paths

[10.4] Renewal processes in quantum mechanics: Excitation of a quantum system by an electromagnetic field generated by a current obtained when electrons strike a detector at renewal times.

Chapter 11:LDP in device physics, quantum scattering amplitudes, quantum filtering and quantum antennas

[11.1] **Large deviations in vacuum polarization**

[11.2] An application of the EKF and LDP to estimating the current in a pn junction The basic equations governing current in a pn semiconductor junction are the diffusion current equation that expresses the current as proportional to the concentration gradient of the minority carriers, the drift current equation, namely Ohm's law, Poisson's equation that relates minority charge concentration to the electric potential or equivalently Gauss' law that relates minority charge concentration to the electric field, the charge conservation equation or equivalently the equation of continuity and finally random perturbations to these equations caused by carriers generating thermal noise. This set of equations thus constitutes a system of stochastic partial differential equations in time and one space dimension. By taking measurements on the current at the end terminals of the junction, we can apply the EKF to estimate the current within the semiconductor. Finally, the difference between the true stochastic current field and its EKF estimate will satisfy approximately a linearized stochastic pde driven by weak thermal noise. Shot noise terms may also be included in this system of equations and hence an LDP rate function for this error can be derived.

[11.3] Large deviation problems in quantum stochastic filtering theory:

[11.4] Large deviation problems in quantum antennas. The Maxwell and Dirac field equations are set up taking into account their interactions, ie, the Dirac current source appearing in the Maxwell equations and the interaction of the Dirac wave field with the electromagnetic four potential appearing in the Dirac equation. The Dyson-Schwinger equations for the exact electron and photon propagator equations are set up. Then small amplitude random classical currents and electromagnetic fields acting as perturbing sources to these equations are also considered and large deviation rate functions for the propagator are obtained. More generally, quantum stochastic current and field sources can also be considered and then higher moments of the Dirac wave field and the photon field in the tensor product of Bosonic and Fermionic coherent states of the quantum field and the quantum noise sources can be evaluated using higher order perturbation theory. These expressions give us the far field quantum statistical moments of the photon and electron radiation fields. The joint treatment of photons, electrons and positrons enables us to speak simulatneously of electron-positron radiation just as we talk about electromagnetic radiation. When the perturbing stochastic and quantum stochastic fields are of weak amplitude, then we can even talk about the LDP rate functions of the probability distribution of a single observable expressible as a functional of the Dirac and photon wave fields.

Chapter 12:How the electron acquires its mass, estimating the electron spin and the quantum electromagnetic field within a cavity in the presence of quantum noise

[12.1] Large deviation methods in classical and quantum field theory.

[1] Corrections to the electron mass from electro-weak interactions.

[2] The role of gravity in giving extra mass to the electron.

[3] Estimating the electromagnetic field at a given spatial point within the cavity as well as the spin of the electron bound to its nucleus placed within the cavity.

Chapter 13: Mathematical tools for large deviations, neural networks, LDP in physical theories, EM and LDP algorithms in quantum parameter estimation and filtering

[13.1] The Ascoli-Arzela theorem as a condition on a family of functions on a metric space to form a compact set, The Stone-Weierstrass theorem applied to prove separability of the family of continuous functions on a compact metric space, Application of the Stone-Weierstrass theorem to prove Prohorov's tightness theorem as an equivalent condition for compactness of a set of probability measures on a separable metric space, application of the Ascoli-Arzela theorem and Prohorov's tightness theorem for obtaining a condition for a set of probability measures on the space of continuous functions on a compact real interval to be compact (ie, have a convergent subsequence in the weak topology), application of this result to the proof of the invariance principle of Donsker, namely convergence of discrete random walks with interpolation to Brownian motion.

[13.2] The Prohorov tightness theorem

[13.3] Some remarks on neural networks related to large deviation theory.

[13.4] LDP problems in general relativity and non-Abelian gauge field theory. Expressing the non-Abelian antisymmetric gauge field in terms of the gauge potential and the curvature tensor of space-time in terms of the gravitational connection using the language of differential forms on Lie algebras and hence to determine the statistics of the field and curvature fluctuations when there is a small fluctuation in the potential and the connection.

[13.5] Large deviations in String theoretic corrections to field theories. The quantum string is a particle plus a small quantum fluctuation around this point particle with the quantum fluctuation being described as superpositions of string creation and annihilation operators and the associated coefficients being functions of the string length parameter. The Lagrangian of a field evaluated at this quantum string integrated over the string length parameter can be expressed as the sum of a point field Lagrangian plus small quantum fluctuation terms that are superpositions of polynomials in the creation and annihilation operators. The statistics, ie probability distribution of the quantum fluctuation part of this Lagrangian in a coherent state of the string can in principle be computed and a large deviation rate function for this statistics can be evaluated in the limit as the small parameter in the the string quantum fluctuation converges to zero.

Chapter 14: Quantum transmission lines, engineering applications of stochastic processes

[14.1] Summary of the theory of quantum transmission lines using the GKSL theory of open quantum systems.

[14.2] Kolmogorov's existence for stochastic processes applied to the problem of describing infinite image fields, ie, image fields with a countably infinite number of pixels

[14.3] Dirichlet series with image processing applications

[14.4] Linear, nonlinear and stochastic phenomena in classical and quantum transmission line and waveguide theory.

[14.5] An application of the EM algorithm to a quantum parameter estimation problem and to quantum filtering theory.

[a] Parameters of the atomic system like electronic charge, mass and number of electrons are to be estimated by taking measurements using a PVM on the quantum mixed state at a succession of times taking into account the state collapse postulate following each measurement and also the distribution of the latent random variable which parametrizes a classical random electromagnetic field incident upon the atom. The joint measurement probabilities at this succession of times conditioned upon the latent random variable is computed in the usual way ie taking the collapse postulate into account at each measurement time point followed by Schrodinger evolution of the state in the time interval between any two successive measurement times. The EM algorithm is then applied to maximizing this joint probability averaged over the latent random variable.

[b] When the Lindlbad noise operator parameters in the Hudson-Parthasarathy-Schrodinger quantum stochastic differential equation become very small and non-demolition measurements are used to construct the corresponding Belavkin filter, then what is the rate function of the Belavkin filter state estimate process over a given time interval compared with the true state process of the noiseless Schrodinger equation ?

[14.6] The EM algorithm and large deviation theory: Evaluating the rate function of the error between the successive iterated EM parameter estimate and the true parameter value in the limit as the number of iid measurements becomes very large.

[14.7] Kolmogorov-Smirnov statistics for computing the probability that the empirical distribution will stay within a boundary around the true distribution of a sequence of iid random variables in the limit as the number of measurements becomes very large. This computation is based on convergence of random walk to Brownian motion and application of the strong Markov property of Brownian motion to the stop times at which the Brownian motion process successively hits a boundary.

[14.8] Quantum transmission lines, LDP problems. Using the Glauber-Sudarshan non-orthogonal resolution of operators in terms of integrals over coherent states, derive the pde satisfied by the density operator representation for the quantum transmission line after noting that the line differential equations can be derived using the Heisenberg-Lindblad matrix mechanics with harmonic oscillator Hamiltonian and Lindblad operators being represented as linear combinations of the harmonic oscillator creation and annihilation operators. The Lindblad noise operators account for quantum noise fluctuations and dissipation introduced by the bath. Derive the density evolution from unitary dynamics for system and bath by partial tracing over the bath using the standard Hudson-Parthasarathy formalism. Finally, in the limit as the Lindblad operators become very small, derive the LDP rate function for the Belavkin filter state estimate of the line based on non-demolition measurements. The measurement should correspond to noisy measurements of some functional of the line voltage and current.

Chapter 15:More tools in probability, electron mass in the presence of gravity and electromagnetic radiation, more on LDP in quantum field theory, non-Abelian gauge field theory and gravitation

1. Weak convergence of probability measures on a metric space
2. The Lindberg conditions and the central limit theorem
3. The Stone-Weierstrass theorem and its application to the proof of Prohorov's compactness theorem

[15.1] The Stone-Weierstrass theorem.

The notion of weak convergence of probability measures on a metric space is a far reaching generalization of the notion of convergence of a family of probability distributions on a finite dimensional vector space to another distribution. This notion enables us in particular to construct a sequence of stochastic processes on a time interval and prove that the finite dimensional distributions of such a process converge weakly and then prove that the probability measures induced by the stochastic processes are tight and hence by Prohorov's theorem deduce that they are compact and hence every infinite subset of these probability measures has a weakly convergent subsequence with the limiting measures all having the same finite dimensional distributions. But if all the finite dimensional distributions of a probability measure on a vector space are known, the measure itself is uniquely determined. This method thereby enables us to prove invariance principles like convergence of random walks to Brownian motion and more generally to diffusion processes. It also as in the Kolmogorov Centsov theorem enables us to construct the Brownian motion process as a process with continuous sample paths having the prescribed finite dimensional Gaussian distribution.

[15.2] On the amount of mass that an electron can get from the background electromagnetic and gravitational fields

[15.3] Electron propagator corrections in the presence of quantum noise.

[1] Dirac Hamiltonian in a radial potential, application of Large deviation theory to computing the statistics of the quantum average of an observable in the presence of Hudson-Parthasarathy noise; corrections to the electron and photon propagator in the presence of Hudson-Parthasarathy noise.
[15.4] Large deviation problems for the Schrodinger and Dirac noisy channels.

[15.5] Central limit theorem for martingales

[15.6] More problems in LDP applied to quantum field theory. Analogy between the the rate function in classical probability theory and the quantum effective action as the Legendre transform of the logarithm of the path integral for field in the presence of an external classical current source field. Deriving the equations of motion for the average of the quantum field in terms of the classical current source and vice versa using the quantum effective action.

[15.7] Schrodinger and Klein-Gordon equations in quantum field theory based on an infinite dimensional Laplacian operator.

[15.8] Proof of the Prohorov tightness theorem.

[15.9] Large deviation problems in field measurement analysis

[15.10] ADM action for quantum gravity and its noisy perturbation with LDP analysis of the solution metric.

[15.11] The Bianchi identity for non-Abelian gauge fields.

[15.12] More problems in LDP applied to quantum field theory.

[15.13] Schrodinger and Klein-Gordon equations in quantum field theory based on an infinite dimensional Laplacian operator.

Chapter 16: Weak convergence, Sanov's theorem, LDP in binary signal detection
[16.1] Prohorov's tightness theorem, "necessity part".

[16.2] Sanov's theorem for discrete random variables.
[16.3] LDP in binary phase shift keying.

[16.4] Some problems related to probability distributions on compact metric spaces.

[16.5] Large deviations for frequency modulated signals

Chapter 17: LDP for classical and quantum transmission lines, string theoretic corrections to classical field Lagrangians, non-Abelian gauge theory in the language of differential forms

[17.1] LDP theory applied to transmission lines with line loading.

[17.2] LDP formulation of the quantum transmission line. Derive the Line equations from Heisenberg-Lindblad matrix mechanics for open quantum systems with harmonic oscillator Hamiltonian. Derive the this Heisenberg-Lindlbad equations from the Hudson-Parthasarathy-Schrodinger equation for open quantum systems by introducing creation and annihilation processes. Derive the Belavkin quantum filter equations for the filtered state from non-demolition measurements involving measurement of some linear combination of the spatial Fourier series components of the line voltage and current. Introduce small parameters into the Lindblad noise operators and hence derive the rate functional for the filtered density operator in a coherent state. This will exist because the Belavkin equations is a classical stochastic Schrodinger equation for the density operator.

[17.3] Yang-Mills gauge fields, The Euler characteristic, string theoretic corrections to the Yang-Mills anomaly cancellation Lagrangian terms. When the gauge potential and metric tensor have small random fluctuations, then how does one compute the rate function of the corresponding gauge and gravitational anomaly correction terms in the super-gravity plus super-Yang-Mills Lagrangian.

[17.4] Quantum averaging based derivation of action functional for point fields from action functional of string fields.

Chapter 18: LDP and EM algorithm, LDP for parameter estimates in linear dynamical systems, philosophical questions in quantum general relativity, Dirac operator for Yang-Mills field in curved space-time, propagator integral representation for general quantum fields

[18.1] Large deviations and the EM algorithm: Large deviation properties of parameter estimates derived using the EM algorithm in the presence of noise relative to ML parameter estimates obtained in the absence of noise when there are latent random parameter vectors in the measurement model.

[18.2] Fundamental problems in quantum general relativity, the problem of defining time as a non-commutative observable. When the time observable is a scalar plus a small non-commutative observable, we can pose large deviations problems like how does the statistics of an observable at this non-commutative time instant behave relative to its value at the scalar component of this non-commutative observable in the limit when the non-commutative component of the time observable converges to zero ? By this we mean what is the rate function of the family of probability distributions of the difference between the observable at the scalar time component and at the non-commutative component when the latter is scaled by a parameter that converges to zero?

[18.3] A problem in Large deviations and Lie algebras. When the parameters of a linear dynamical system undergo small random fluctuations with a given rate function, then determine the rate function of the state process of the dynamical system using Lie algebraic representations of the exponential function of the sum of two matrices.

[18.4] Quantum gravity using holonomy fields.

[18.5] Square of the Dirac operator in curved space-time in the presence of non-Abelian connections.

[18.6] Exponential equivalence, The Dawson-Gartner theorem on LDP for projective limits.

[18.7] Applications

[18.8] Equivalence of LDP rate functions for exponentially equivalent random families.

[18.9] Lehmann's representation of the general propagator as an integral over single particle propagators with varying masses.

For Product Safety Concerns and Information please contact our EU
representative GPSR@taylorandfrancis.com
Taylor & Francis Verlag GmbH, Kaufingerstraße 24, 80331 München, Germany

www.ingramcontent.com/pod-product-compliance
Ingram Content Group UK Ltd.
Pitfield, Milton Keynes, MK11 3LW, UK
UKHW022316250926
475702UK00046B/880
* 9 7 8 1 0 3 2 4 2 5 4 7 4 *